The Hunt for Zero Point

The Hunt for Zero Point

One Man's Journey to Discover the Biggest Secret
Since the Invention of the Atom Bomb

Nick Cook

CENTURY · LONDON

Published by Century in 2001

1 3 5 7 9 10 8 6 4 2

First published in the United Kingdom in 2001 by Century
The Random House Group Limited
20 Vauxhall Bridge Road, London SW1V 2SA

Random House Australia (Pty) Limited
20 Alfred Street, Milsons Point, Sydney,
New South Wales 2061, Australia

Random House New Zealand Limited
18 Poland Road, Glenfield,
Auckland 10, New Zealand

Random House South Africa (Pty) Limited
Endulini, 5a Jubilee Road, Parktown 2193, South Africa

The Random House Group Limited Reg. No. 954009

www.randomhouse.co.uk

A CIP catalogue record for this book
is available from the British Library

ISBN 0 7126 69531

Typeset in Ehrhardt by MATS, Southend-on-Sea, Essex
Printed and bound in the United Kingdom by
Mackays of Chatham PLC, Chatham, Kent

In memory of Julian Cook, inventor, and Harry Hawker, fighter pilot

Per ardua ad astra

And for my children, Lucy and William, that one day they or their children may see the stars more closely

Author's Note and Acknowledgements

Not everyone who has assisted me in *The Hunt For Zero Point* has wished to be identified.

There are four people – 'Amelia Lopez', 'Lawrence Cross', 'Daniella Abelman' and 'Dr Dan Marckus' – whose identities I have deliberately blurred. They, of course, know who they are and I gratefully acknowledge their help.

It goes without saying that I could not have interpreted the mass of raw science data in this book without the exceptional skills of Dan Marckus. His knowledge, wisdom, humour and friendship have been an inspiration to me.

Everyone else is exactly as identified in the text.

The people and institutions I can thank freely are listed below:

My colleagues at *Jane's Defence Weekly* and *Interavia*, Tom Valone of the Integrity Research Institute, Paul LaViolette Ph.D, Dr Kathleen Bunten, Isabel Best, Joel Carpenter, the Imperial War Museum, Philip Henshall, Graham Ennis, Rowland White, Guy Norris, Jeremy Bartlett, Michael Collins, Guy Rigby, Kevin Bexley, Stephen Parker, Bill Zuk, Yvonne Kinkaid, the Office of (US) Air Force History, David Windle, Chris Gibson, Tony Frost, the late Prof. Brian Young, Dr Ron Evans, BAE Systems, Garry Lyles, George Schmidt, Marc Millis, the US National Aeronautics and Space Administration, Lockheed Martin and the Lockheed Martin Skunk Works, Dr Evgeny Podkletnov, Dr Hal Puthoff, the late Ben Rich, Bill Sweetman, Glenn Campbell, Mark Farmer and the Groom Lake Interceptors, Mike Grigsby, James Cook, David Monaghan, George Muellner, Marian Swedija, Irving Baum, Ken Jones, Charles T. Schaeffer, the 16th Armored Division veterans' association, Dr Amy Schmidt, the US National Archives Records Administration, Ned Bedessem, US Army Center of Military History, the Bundesarchiv at Freibourg, the UK Public Records Office, Maurizio Verga, James Holland, Callum Coats, Don Kelly, Bob Widmer, Boyd Bushman, George Hathaway, John Hutchison and John B. Alexander.

My particular thanks go to:

Mark Booth, Hannah Black, Anna Cherrett and the whole team at Century, Bill Rose for his outstanding picture research, Igor Witkowski for taking me to the Wenceslas Mine and for sharing his tremendous insight into German secret weapons, Tom Agoston for his ground-breaking book *Blunder!*, the Schauberger family – Frau Ingeborg, Joerg and Ingrid – for their kindness and hospitality at Bad Ischl, Brue Richardson for her constancy, serenity and guidance, Barry and Fiona Shaw for putting up with me (and the rest) during a myriad of writing weekends and Mark Lucas for being a damn fine friend and agent.

Last, but not least, I would like to thank Ali, my family and hers. This book would have been quite impossible without them.

Prologue

The dust devils swirled around my Chevrolet Blazer, catching the early evening light. I watched Sheriff's Deputy Amelia Lopez clamber out of her Chrysler Le Baron and stare for a moment in the direction the plane had gone down ten years ago.

I grabbed my rucksack. By the time I looked up again she was already striding towards the peak.

Over a rusted barbed-wire fence and we were into the scrub – with new traces of green at its tips from the spring rains. Beyond lay the edge of the Sequoia National Forest, a huge expanse of protected park and woodland.

We left the broken fence posts behind and our cars were lost against the sunset. I looked for other traces of a human presence, but found none, even though we were only twelve miles from Bakersfield, California, a city of four hundred thousand people on the edge of the Sierra Nevadas.

Amelia Lopez' peaked cap and the firearm on her gunbelt were clearly silhouetted as she moved along the jagged edge of the ridgeline.

As I sucked down the warm thin air and wiped the sweat out of my eyes, I tried to picture her as she must have been on that sweltering July night ten years earlier; the night she'd been out partying with her college friends at a campsite near the Kern River.

It was in the breaking hours of 11 July 1986. Just as she was settling into her sleeping bag the jet went supersonic somewhere in the black sky overhead.

The pressure wave of the sonic boom hit the campsite like a clap of thunder, sending a shower of embers from the camp fire into the night sky.

Amelia was too startled to say a thing; then, the entire horizon was flash-lit by an enormous explosion, the flames shooting skywards as the plane ploughed into Saturday Peak ten miles away.

She told me it had sparked a dozen brush fires on the edge of the forest; that it took more than a hundred Forest Service and local firefighters to put it out. Her only thought was that this wasn't an aircraft at all, but a hydrogen bomb.

Within hours, every newspaper in the state, and a whole lot more besides, had a reporter heading for these foothills with a brief to find out what had hit the ground. Amelia Lopez, a law student at the state university in Sacramento, had been one of several witnesses quoted in the papers that had covered the story, which was how I'd traced her.

She and her friends hadn't gone more than five miles towards the impact point when one of them noticed a figure on the trail up ahead. The scrub either side of her erupted with movement and the next thing she knew her face was in the dirt and she had a boot in her back and a gun at her head.

Out of the corner of her eye she saw that they were soldiers – not California National Guard, as might have been expected in an environmental emergency, but SWAT-types brandishing assault rifles, night-vision systems and a shit-load of threats about government property and national security.

Two of her friends started on about their rights under the Constitution, that this was public land and there wasn't a person on earth who could tell them to get off it. But to Lopez their protests registered as white noise on the edge of a persistent and piercing alarm. These soldiers were unlike any she'd ever seen.

She screamed for her friends to shut up, but once the screaming began, she couldn't stop it. She screamed and she yelled and she flailed against the pressure in her back, until the next thing she felt was the slap from her room-mate that brought her round. When she finally understood what she was being told, it was that the soldiers were gone.

None of them said a word as they doubled back to the campsite. When they reached it, still numb with shock, they found a bunch of reporters on-site getting statements from other witnesses. Somebody shoved a tape-recorder in her face and started asking questions and before she knew it she'd given her name and stammered something about an atomic blast.

As for the rest, she and her friends said nothing.

Amelia Lopez kept a lid on her feelings for the next two and half years, till November 1988, in fact, when the outgoing Reagan administration revealed the existence of the F-117A Stealth Fighter, an aircraft that had been flying in secret squadron service out of a classified air base in Nevada for over five years. In that time, she learned, it had crashed twice: and on one of those occasions – on the night of 10/11 July 1986, to be precise – she'd had the grave misfortune to be there.

The troops had been part of a Pentagon 'red team' flown in by helicopter to secure the crash site at all costs.

When she got my message, she'd been reluctant to meet up at first, but when I finally persuaded her to talk, she found it difficult to stop.

We reached the crash site soon after the sun dipped below the edge of the mountain. The summit was only 2,000 feet above us, but here the ground was even and covered in a crusty layer of dirt. The plants and trees were younger than the vegetation we'd passed on the way up. But that was the only real clue something had happened here.

Amelia Lopez sat on a rock and slowly removed her mirror shades before pouring bottled springwater over her face. I felt her eyes follow me as I moved between clumps of vegetation, kicking over rocks and sifting the sand, even though there was nothing to see.

Lopez bent down and ran her fingers through the soil. 'I read they sieved the dirt for a thousand yards out from the impact point,' she said. 'Those guys were damned thorough. A few weeks after they left it was like *nothing* ever happened here.' She paused a moment before adding: 'You being an expert, I imagine you knew that.'

It was framed as a question and I grappled for something to say, conscious that she'd brought me here for any insight I could provide into the events of that night.

I said nothing, so she turned to me and said: 'Are you gonna tell me what is really going on here?'

Overhead, an eagle cried. As I watched it wheel on the updrafts I hoped that she wouldn't press me for an answer, because I didn't know what to tell her.

Standing here in this place, I was filled with the old feeling. It was almost impossible to articulate, but it left you with a taste in the mouth, some innate sense, that however far you dug, however many people you interviewed or questioned, you were simply scratching the surface of the sprawling US defence-industrial base. What had happened here, the events that had imprinted themselves onto the landscape in a moment or two of madness a decade or so earlier, were almost tangible, even though there was no physical evidence – no fragments amidst the thin soil and the rocks – to suggest anything out of the ordinary had occurred.

These people were thorough; Lopez herself had said it. But they left something behind, something you couldn't see or touch – and it was that trace, that echo of past deeds, that had brought me here.

The Stealth Fighter was real enough. As a reporter, I'd covered it from the inside out. Yet as a piece of technology it was more than two decades old, almost every detail of it in the open now. But strip away the facts and the feeling persisted.

I got it when I went to US government defence laboratories and on empty wind-blown hangar floors in parched, little-known corners of the country. I got it at press conferences in power-soaked corridors of the Pentagon. But most of all I got it when I stared into the eyes of the people who worked on those programmes.

What I got back was a look. Individually, it said nothing, but collectively it told me there was a secret out there and that it was so big no one person held all the pieces. I knew, too, that whatever it was, the secret had a dark heart, because I could smell the fear that held it in place.

It was impossible to tell Lopez any of this, of course, because it was simply a feeling. But as I headed back for the car, I knew the trip hadn't been a waste.

At long last, the secret had an outline.

Through half-closed eyes, I could almost reach out and touch it.

Chapter 1

From the heavy-handed style of the prose and the faint hand-written '1956' scrawled in pencil along the top of the first page, the photocopied pages had obviously come from some long-forgotten schlock popular science journal.

I had stepped away from my desk only for a few moments and somehow in the interim the article had appeared. The headline ran: 'The G-Engines Are Coming!'

I glanced around the office, wondering who had put it there and if this was someone's idea of a joke. The copier had cut off the top of the first page and the title of the publication with it, but it was the drawing above the headline that was the giveaway. It depicted an aircraft, if you could call it that, hovering a few feet above a dry lakebed, a ladder extending from the fuselage and a crewmember making his way down the steps dressed in a US-style flight-suit and flying helmet – standard garb for that era. The aircraft had no wings and no visible means of propulsion.

I gave the office another quick scan. The magazine's operations were set on the first floor. The whole building was open-plan. To my left, the business editor was head-down over a proof-page checking copy. To her right was the naval editor, a guy who was good for a wind-up, but who was currently deep into a phone conversation and looked like he had been for hours.

I was reminded of a technology piece I'd penned a couple of years earlier about the search for scientific breakthroughs in US aerospace and defence research. In a journal not noted for its exploration of the fringes of paranormality, nor for its humour, I'd inserted a tongue-in-cheek reference to gravity – or rather to anti-gravity, a subject beloved of science-fiction writers.

'For some US aerospace engineers,' I'd said, 'an anti-gravity propulsion system remains the ultimate quantum leap in aircraft design.' The implication was that anti-gravity was the aerospace equivalent of the holy grail: something longed for, dreamed about, but beyond reach – and likely always to remain so.

1

Somehow the reference had escaped the sub-editors and, as a result, amongst my peers, other aerospace and defence writers on the circuit, I'd taken some flak for it. For Jane's, the publishing empire founded on one man's obsession with the detailed specifications of ships and aircraft almost a century earlier, technology wasn't something you joked about.

The magazine I wrote – and still write – for, *Jane's Defence Weekly*, documented the day-to-day dealings of the multi-billion dollar defence business. *JDW*, as we called it, was but one of a portfolio of products detailing the ins and outs of the global aerospace and defence industry. If you wanted to know about the thrust-to-weight ratio of a Chinese combat aircraft engine or the pulse repetition frequency of a particular radar system, somewhere in the Jane's portfolio of products there was a publication that had the answers. In short, Jane's was, and always has been, about facts. Its motto was, and still is: Authoritative, Accurate, Impartial.

It was a huge commercial intelligence-gathering operation; and provided they had the money, anyone could buy into its vast knowledge-base.

I cast a glance at the bank of sub-editors' work-stations over in the far corner of the office, but nobody appeared remotely interested in what was happening at my desk. If the subs had nothing to do with it, and usually they were the first to know about a piece of piss-taking that was going down in the office, I figured whoever had put it there was from one of the dozens of other departments in the building and on a different floor. Perhaps my anonymous benefactor had felt embarrassed about passing it on to me?

I studied the piece again.

The strapline below the headline proclaimed: 'By far the most potent source of energy is gravity. Using it as power future aircraft will attain the speed of light.' It was written by one Michael Gladych and began: 'Nuclear-powered aircraft are yet to be built, but there are research projects already under way that will make the super-planes obsolete before they are test-flown. For in the United States and Canada, research centers, scientists, designers and engineers are perfecting a way to control gravity – a force infinitely more powerful than the mighty atom. The result of their labors will be anti-gravity engines working without fuel – weightless airliners and space ships able to travel at 170,000 miles per second.'

On any other day, that would have been the moment I'd have consigned it for recycling. But something in the following paragraph caught my eye.

The G-Engines Are Coming!

By far the most potent source of energy is gravity. Using it as power future aircraft will attain the speed of light.

By MICHAEL GLADYCH

■ Nuclear-powered aircraft are yet to be built, but there are research projects already under way that will make the super-planes obsolete before they are test-flown. For in the United States and Canada research centers, scientists, designers and engineers are perfecting a

which there has been no escape. "What goes up must come down," they said. The bigger the body the stronger the gravity attraction it has for other objects . . . the larger the distance between the objects, the lesser the gravity pull. Defining those rigid rules was as

way to earth.
This discovery gave modern scientists a new hope. We already knew how to make magnets by coiling a wire around an iron core. Electric current running through the coiled wire created a magnetic field and it could be switched on

The gravity research, it said, had been supported by the Glenn L. Martin Aircraft Company, Bell Aircraft, Lear 'and several other American aircraft manufacturers who would not spend millions of dollars on science fiction.' It quoted Lawrence D. Bell, the founder of the plane-maker that was first to beat the sound barrier. 'We're already working on nuclear fuels and equipment to cancel out gravity.' George S. Trimble, head of Advanced Programs and 'Vice President in charge of the G-Project at Martin Aircraft', added that the conquest of gravity 'could be done in about the time it took to build the first atom bomb.'

A little further on, it quoted 'William P. Lear, the chairman of Lear Inc., makers of autopilots and other electronic controls.' It would be another decade before Bill Lear went on to design and build the first of the sleek business jets that still carry his name. But in 1956, according to Gladych, Lear had his mind on other things.

'All matter within the ship would be influenced by the ship's gravitation only,' Lear apparently said of the wondrous G-craft. 'This way, no matter how fast you accelerated or changed course, your body would not feel it any more than it now feels the tremendous speed and acceleration of the earth.' The G-ship, Gladych explained, could take off like a cannon-shell, come to a stop with equal abruptness and the passengers wouldn't even need seat-belts. This ability to accelerate rapidly, the author continued, would make it ideal as a space vehicle capable of acceleration to a speed approaching that of light.

There were some oblique references to Einstein, some highly dubious 'facts' about the nature of sub-atomic physics and some speculation about how various kinds of 'anti-gravity engines' might work.

But the one thing I kept returning to were those quotes. Had Gladych made them up or had Lawrence Bell, George S. Trimble and William 'Bill' Lear really said what he had quoted them as saying?

Outside, the rain beat against the double-glazed windows, drowning the sound of the traffic that crawled along the London to Brighton road and the unrelenting hum of the air conditioning that regulated the temperature inside.

The office was located in the last suburb of the Greater London metropolis; next stop the congested joys of the M25 ring-road and the M23 to Gatwick Airport. The building was a vast red brick two-storey bunker amid between-the-wars grey brickwork and pebbledash. The rain acted like a muslin filter, washing out what little ambient colour Coulsdon possessed. In the rain, it was easy to imagine that nothing much had changed here for decades.

As aviation editor of *JDW*, my beat was global and it was pretty much unstructured. If I needed to cover the latest air-to-surface weapons developments in the USA, I could do it, with relatively few questions asked. My editor, an old pro, with a history as long as your arm in publishing, gave each of us, the so-called 'specialists' (the aviation, naval and land systems editors), plenty of rope. His only proviso was that we filed our expenses within two of weeks of travel and that we gave him good, exclusive stories. If I wanted to cover an aerospace and defence exhibition in Moscow, Singapore or Dubai, the funds to do so were almost always there.

As for the job itself, it was a mixture of hard-edged reporting and basic provision of information. We reported on the defence industry, but we were part of it, too – the vast majority of the company's revenue coming from the same people we wrote about. Kowtowing was a no-no, but so was kicking down doors. If you knew the rules and played by them you could access almost any part of the global defence-industrial complex. In the course of a decade, I'd visited secret Russian defence facilities and ultra-sensitive US government labs. If you liked technology, a bit of skulduggery and people, it was a career made in heaven. At least 60 per cent of the time I was on the road. The bit I liked least was office downtime.

Again, I looked around for signs that I was being set up. Then, satisfied that I wasn't, but feeling self-conscious nonetheless, I tucked the Gladych article into a drawer and got on with the business of the day. Another aerospace and defence company had fallen prey to post-Cold War economics. It was 24 hours before the paper closed for press and the news editor was yelling for copy.

Two days later, in a much quieter moment, I visited the Jane's library. It was empty but for the librarian, a nice man way past retirement age who used to listen to the BBC's radio lunchtime news whilst gazing out over the building's bleak rear lot.

In the days before the Internet revolution, the library was an invaluable resource. Fred T. Jane published his first yearbook, *Jane's Fighting Ships*, in 1898; and in 1909 the second, *Jane's All The World's Aircraft*, quickly built on the reputation of the former as a reference work *par excellence* for any and all information on aeronautical developments. Nigh on a century later, the library held just about every book and magazine ever put out by the company and a pile of other reference works besides.

I scanned the shelves till I found what I was looking for.

The *Jane's All The World's Aircraft* yearbook for 1956 carried no mention of anti-gravity experiments, nor did successive volumes, but that came as no great surprise. The yearbooks are the aerospace equivalent of *Burke's Peerage* or the *Guinness Book of Records*: every word pored over, analysed and double-checked for accuracy. They'd have given anti-gravity a very wide berth.

For a story like this, what I was looking for was a news publication.

I looked along the shelves again. Jane's had got into the magazine publishing business relatively recently and the company's copies of *Flight International* and *Aviation Week* only ran back a few years. But it did have bound volumes of *Interavia Aerospace Review* from before the Second World War. And it was on page 373 in the May 1956 edition of this well respected publication, in amongst advertisements for Constellation airliners, chunky-looking bits of radar equipment and (curiously for an aviation journal) huge 'portable' Olivetti typewriters, that I found a feature bylined 'Intel, Washington D.C.' with the headline: 'Without Stress or Strain . . . or Weight.' Beneath it ran the strapline: 'The following article is by an American journalist who has long taken a keen interest in questions of theoretical physics and has been recommended to the Editors as having close connections with scientific circles in the United States. The subject is one of immediate interest, and Interavia would welcome further comment from knowledgeable sources.'

The article referred to something called 'electro-gravitics' research, whose aim was to 'seek the source of gravity and its control.' This research, 'Intel' stated, had 'reached a stage where profound implications for the entire human race are beginning to emerge.'

I read on, amused by the tone and wondering how on earth the article had come to be accepted in a mainstream aerospace journal.

'In the still short life of the turbo-jet airplane [by then, 1956, little more than a decade], man has had to increase power in the form of brute thrust some twenty times in order to achieve just twice the speed. The cost in money in reaching this point has been prodigious. The cost in highly specialised man-hours is even greater. By his present methods man actually fights in direct combat the forces that resist his efforts. In conquering gravity he would be putting one of his most competent adversaries to work for him. Anti-gravitics is the method of the picklock rather than the sledge-hammer.'

Not only that, the article stated, but anti-gravity could be put to work in other fields beyond aerospace. 'In road cars, trains and boats the headaches of transmission of power from the engine to wheels or propellers would simply cease to exist. Construction of bridges and big buildings would be greatly simplified by temporary induced weightlessness etc. Other facets of work now under way indicate the possibility of close controls over the growth of plant life; new therapeutic techniques, permanent fuelless heating units for homes and industrial establishments; new sources of industrial power; new manufacturing techniques; a whole field of new chemistry. The list is endless . . . and growing.'

It was also sheer fantasy.

Yet, for the second time in a week I had found an article – this time certainly in a publication with a solid reputation – that stated that US aerospace companies were engaged in the study of this 'science'. It cited the same firms mentioned by Gladych and some new ones as well: Sperry-Rand and General Electric among them. Within these institutions, we were supposed to believe, people were working on theories that could not only make materials weightless, but could actually give them 'negative weight' – a repulsive force that would allow them to loft away 'contra-gravitationally'. The article went further. It claimed that in experimentation conducted by a certain 'Townsend T. Brown' weights of some materials had already been cut by as much as 30 per cent by 'energizing' them and that model 'disc airfoils' utilising this technology had been run in a wind-tunnel under a charge of a hundred and fifty kilovolts 'with results so impressive as to be highly classified'.

I gazed out over the slate rooftops. For *Interavia* to have written about anti-gravity, there had to have been something in it. The trouble was, it was history. My bread-and-butter beat was the aerospace industry of the 1990s, not this distant cosy world of the fifties with its heady whiff of jet-engine spirit and the developing Cold War.

I replaced the volume and returned to my desk. It should have been easy to let go, but it wasn't. If people of the calibre quoted by Gladych and *Interavia* had started talking about anti-gravity anytime in the past ten years I would have reported it – however sceptical I might be on a personal level. Why had these people said the things they had with such conviction? One of them, George S. Trimble, had gone so far as to predict that a breakthrough would occur in around the same time it took to develop the atomic bomb – roughly five years. Yet, it had never happened. And even if the results of 'Townsend T. Brown's' experiments had been 'so impressive as to be highly classified' they had clearly come to naught; otherwise, by the 60s or 70s the industry would have been overtaken by fuelless propulsion technology.

I rang a public relations contact at Lockheed Martin, the US aerospace and defence giant, to see if I could get anything on the individuals Gladych had quoted. I knew that Lawrence Bell and Bill Lear were both dead. But what about George S. Trimble? If Trimble was alive – and it was a long shot, since he would have to be in his 80s – he would undoubtedly confirm what I felt I knew to be true; that he had been heavily misquoted or that anti-gravity had been the industry's silly-season story of 1956.

A simple phone call would do the trick.

Daniella 'Dani' Abelman was an old media contact within Lockheed Martin's public affairs organisation. Solid, reliable and likeable, she'd grown up in the industry alongside me, only on the other side of the divide. Our relationship with the information managers of the aerospace and defence world was as double-edged as the PR/reporter interface in any other industry. Our job was to get the lowdown on the inside track and, more often than not, it was bad news that sold. But unlike our national newspaper counterparts, trade press hacks have to work within the industry, not outside it. This always added an extra twist to our quest for information. The industry comprised hundreds of thousands of people, but despite its size, it was surprisingly intimate and incestuous enough for everyone to know everyone else. If you pissed off a PR manager in one company, even if it was on the other side of the globe, you wouldn't last long, because word would quickly get around and the flow of information would dry up.

But with Abelman, it was easy. I liked her. We got on. I told her I needed some background on an individual in one of Lockheed Martin's 'heritage' companies, a euphemism for a firm it had long since swallowed whole.

The Glenn L. Martin Company became the Martin Company in 1957. In 1961, it merged with the American-Marietta Company, becoming Martin-Marietta, a huge force in the Cold War US defence electronics industry. In 1994, Martin-Marietta merged again, this time with Lockheed to form Lockheed Martin. The first of the global mega-merged defence behemoths, it built everything from stealth fighters and their guided weapons to space launchers and satellites.

Abelman was naturally suspicious when I told her I needed to trace an ex-company employee, but relaxed when I said that the person I was interested in had been doing his thing more than 40 years ago and was quite likely dead by now.

I was circumspect about the reasons for the approach, knowing full well if I told her the real story, she'd think I'd taken leave of my senses.

But I had a bona fide reason for calling her – and one that legitimately, if at a stretch, involved Trimble: I was preparing a piece on the emergence of the US aerospace industry's 'special projects' facilities in the aftermath of the Cold War.

Most large aerospace and defence companies had a special projects unit; a clandestine adjunct to their main business lines where classified activities could take place. The shining example was the Lockheed Martin 'Skunk Works', a near-legendary aircraft-manufacturing facility on the edge of the California high desert.

For 50 years, the Skunk Works had sifted Lockheed for its most highly skilled engineers, putting them to work on top-secret aircraft projects.

Using this approach it had delivered some of the biggest military breakthroughs of the 20th century, among them the world's first Mach 3 spyplane and stealth, the art of making aircraft 'invisible' to radar and other enemy sensor systems.

But now the Skunk Works was coming out of the shadows and, in the process, giving something back to its parent organisation. Special projects units were renowned for bringing in complex, high-risk defence programmes on time and to cost, a skill that had become highly sought-after by the main body of the company in the austere budget environment of the 1990s.

Trimble, I suggested, might be able to provide me with historical context and 'colour' in an otherwise dry business story. 'Advanced Programs', the outfit he was supposed to have worked for, sounded a lot like Martin's version of the Skunk Works.

Abelman said she'd see what she could do, but I wasn't to expect any short-order miracles. She wasn't the company historian, she said drily, but she'd make a few enquiries and get back to me.

I was surprised when she phoned me a few hours later. Company records, to her surprise – and mine – said that Trimble was alive and enjoying retirement in Arizona. 'Sounds hard as nails, but an amazing guy by all accounts,' she breezed. 'He's kinda mystified why you want to talk to him after all this time, but seems okay with it. Like you said, it's historical, right?'

'Right,' I said.

I asked Abelman, while she was at it, for all the background she had on the man. History or not, I said, trying to keep it light, I liked to be thorough. She was professional enough to sound less than convinced by my new-found interest in the past, but promised she'd do her best. I thanked her, then hung up, feeling happy that I'd done something about it. A few weeks, a month at the outside, the mystery would be resolved and I could go back to my regular beat, case closed.

Outside, another bank of grey storm clouds was rolling in above rooftops that were still slick from the last passing shower.

I picked up my coat and headed for the train station, knowing that somewhere between the office and my flat in central London I was going to get soaked right through.

The initial information came a week later from a search through some old files that I'd buried in a collection of boxes in my basement: a company history of Martin Marietta I'd barely remembered I'd acquired. It told me that in 1955 Trimble had become involved in something called the Research Institute for Advanced Studies, RIAS, a Martin spin-off organisation whose brief was to 'observe the phenomena of nature . . . to discover fundamental laws . . . and to evolve new technical concepts for the improvement and welfare of mankind.'

Aside from the philanthropic tone, a couple of things struck me as fishy about the RIAS. First off, its name was as bland as the carefully chosen 'Advanced Development Projects' – the official title – of the Skunk Works. Second, was the nature and calibre of its recruits. These, according to the company history, were 'world class contributors in mathematics, physics, biology and materials science'.

Soon afterwards, I received a package of requested information from Lockheed Martin in the post. RIAS no longer existed, having been subsumed by other parts of the Lockheed Martin empire. But through an old RIAS history, a brochure published in 1980 to celebrate the organisation's 'first 25 years', I was able to glean a little more about Trimble and the outfit he'd inspired. It described him as 'one of the most creative and imaginative people that ever worked for the Company'.

I read on.

From a nucleus of people that in 1955 met in a conference room at the Martin Company's Middle River plant in Maryland, RIAS soon developed a need for its own space. In 1957, with a staff of about 25 people, it moved to Baltimore City. The initial research programme, the brochure said, was focused on NASA and the agency's stated goal of putting a man on the moon. But that wasn't until 1961.

One obvious question was, what had RIAS been doing in the interim? Mainly maths, by the look of it. Its principal academic was described as an expert in 'topology and non-linear differential equations'.

I hadn't the least idea what that meant.

In 1957, the outfit moved again, this time to a large mansion on the edge of Baltimore, a place chosen for its 'campus-like' atmosphere. Offices were quickly carved from bedrooms and workshops from garages.

It reminded me of accounts I'd read of the shirt-sleeves atmosphere of the early days of the Manhattan Project when Oppenheimer and his team of atom scientists had crunched through the physics of the bomb.

And that was the very same analogy Trimble had used. The conquest of gravity, he'd said, would come in the time it took to build the bomb.

I called a few contacts on the science and engineering side of Lockheed Martin, asking them, in a roundabout kind of way, whether there was, or ever had been, any part of the corporation involved in gravity or 'counter-gravitational' research. After some initial questions on their part as to why I should be interested, which I just about managed to palm off, the answer that came back was a uniform 'no'. Well, almost. There was a guy, one contact told me, a scientist who worked in the combat aircraft division in Fort Worth who would talk eloquently about the mysteries of Nature and the universe to anyone who would listen. He'd also levitate paperclips on his desk. Great character, but a bit of a maverick.

'Paperclips?' I'd asked. 'A maverick scientist levitating paperclips on his desk? At Lockheed Martin? Come *on*.'

My source laughed. If he hadn't known better, he'd have said I was working up a story on anti-gravity.

I made my excuses and signed off. It was mad, possibly dangerous stuff, but it continued to have me intrigued.

I called an old friend who'd gained a degree in applied mathematics. Tentatively, I asked whether topology and non-differential linear equations had any application to the study of gravity.

Of course, he replied. Topology and non-linear equations were the standard methods for calculating gravitational attraction.

I sat back and pieced together what I had. It didn't amount to much, but did it amount to something?

In 1957, George S. Trimble, one of the leading aerospace engineers in the US at that time, a man, it could safely be said, with a background in highly advanced concepts and classified activity, had put together what looked like a special projects team; one with a curious remit.

This, just a year after he started talking about the Golden Age of Anti-Gravity that would sweep through the industry starting in the 1960s.

So, what went wrong?

In its current literature, the stuff pumped out in press releases all the time, the US Air Force constantly talked up the 'vision': where it was going to be in 25 years, how it was going to wage and win future wars and how technology was key.

In 1956, it would have been as curious as I was about the notion of a fuelless propulsion source, one that could deliver phenomenal per-formance gains over a jet; perhaps including the ability to accelerate rapidly, to pull hairpin turns without crushing the pilot and to achieve speeds that defied the imagination. In short, it would have given them something that resembled a UFO.

I rubbed my eyes. The dim pool of light that had illuminated the Lockheed-supplied material on Trimble and RIAS had brought on a nagging pain in the back of my head. The evidence was suggesting that in the mid-50s there had been some kind of breakthrough in the anti-gravity field and for a small window in time people had talked about it freely and openly, believing they were witnessing the dawn of a new era, one that would benefit the whole of mankind.

Then, in 1957, everyone had stopped talking about it. Had the military woken up to what was happening, bringing the clamps down?

Those in the know, outfits like Trimble's that had been at the forefront of the breakthrough, would probably have continued their research, assembling their development teams behind closed doors, ready for the day they could build real hardware.

But of course, it never happened.

It never happened because soon after Trimble, Bell, and Lear made their statements, sanity prevailed. By 1960, it was like the whole episode never took place. Aerospace development continued along its structured, ordered pathway and anti-gravity became one of those taboo subjects that people like me never, ever talked about.

Satisified that everything was back in its place and as it should be, I went to bed.

Somewhere in my head I was still tracking the shrill, faraway sounds of the city when the phone rang. I could tell instantly it was Abelman. Separated by an ocean and five time-zones, I heard the catch in her breathing.

'It's Trimble,' she said. 'The guy just got off the phone to me. Remember how he was fine to do the interview? Well, something's happened. I don't know who this old man is or what he once was, but he told me in no uncertain terms to get off his case. He doesn't want to speak to me and he doesn't want to speak to you, not now, not ever. I don't mind telling you that he sounded scared and I don't like to hear old men scared. It makes me scared. I don't know what you were really working on when you came to me with this, Nick, but let me give you some advice. Stick to what you know about; stick to the damned present. It's better that way for all of us.'

Chapter 2

In 1667, Newton mathematically deduced the nature of gravity, demonstrating that the same force that pulls an apple down to earth also keeps the moon in its orbit and accounts for the revolutions of the planets. But today, we are still thwarted in attempts to measure it with any great precision. In lab experiments carried out since the 1930s, G has consistently defied efforts to be measured to more than a few decimal places.

This was what the reference books told me as I ploughed through a stack of standard science works in the musty, gothic surroundings of the local library.

It was intensive work. Science was something I'd come to associate with the grind of exams. It didn't feel like the beginnings of a journalistic investigation.

I continued to scratch notes. But Newton openly stated that he had no idea what gravity actually was. All he knew was that it had to be caused by something.

The idea that a body may act on another through the vacuum of space over huge distances 'without the mediation of anything else . . . is to me so great an absurdity that I believe no man can ever fall into it. Gravity must be caused . . . but whether this agent be material or immaterial I have left to the consideration of my reader.'

I glanced up. The librarian, who'd been waiting to catch my eye, nodded towards the clock on the wall behind her. I looked around and realised I was the only person in the room. I'd lost track; it was a Saturday and the library closed early.

Outside, the rain had given way to the starlit sky of a passing cold front. I pulled up my collar and started down the street, dodging puddles that shimmered under the streetlights. The anomaly over gravity's measurement and the uncertainties over its causes only served to tell me how incomplete my knowledge of physics was.

I reached the edge of the common. The lights of my home street were faintly visible through the trees. I thought about my late-night call from

Abelman. In the week since her approach to Trimble and his initial favourable response to the idea of an interview, it very much appeared that somebody had got to him. And then I thought about what I had just learned. If we had no real understanding of gravity, how could people say with such certainty that *anti-gravity* could not exist?

In 1990, the US Air Force had been looking at developing a weapon capable of firing a 'plasma bullet' – a doughnut-shaped ring of ionised gas – at 10,000 kilometres per second.

Shiva Star was capable of generating and holding up to 10 megajoules of electrical energy and a potential 10 trillion watts – three times as much as the entire US electricity grid carried in a year. At the time that I visited Shiva, which was located within the USAF's directed energy research laboratory at Kirtland Air Force Base in New Mexico, programme engineers had been readying to fire a plasma bullet sometime in 1995. The purpose of the bullet was to destroy incoming Russian nuclear warheads and, despite some fierce technological challenges, the programme engineers were confident they could do it. But several years later, when I returned to Kirtland, it was like the plasma bullet project never existed. Engineers had difficulty even recalling it.

Officially, it had been terminated on cost grounds. But this made little sense. The programme had been budgeted at $3.6 million per year for five years. Eighteen million bucks to produce a true quantum leap weapon system. Few people I spoke to bought the official version. Some-where along the way, they said, Shiva must have delivered. Somewhere along the line, the programme had gone black.

I coupled this knowledge with what the anti-gravity proponents had been saying in 1956. If you could find a way of shielding objects from the effects of gravity, the military, let alone the economic ramifications would be enormous.

Aircraft propulsion seems to have progressed little in appreciable terms since the advent of the jet engine in the 1940s. Incremental improvements for decades have been of the order of a few percentage points. The fastest aircraft in the world – officially, at least – is still the Lockheed Blackbird, designed in the late 1950s, first flown in 1962 and retired in 1990.

Amongst my peers, there had been speculation since the late 1980s about the existence of a secret replacement for the Blackbird, a mythical plane called Aurora that supposedly flew twice as fast and on the edges of space.

I had no direct evidence of Aurora, but then I'd never gone looking for it either. On balance, though, I felt *something* had been developed.

In 1992, circumstantial evidence of Aurora's existence was strengthened when *Jane's Defence Weekly* carried a detailed sighting of a massive triangular-shaped aircraft spotted in formation with USAF F-111 bombers and an air-refuelling tanker above an oil rig in the North Sea. The sighting was credible because it was witnessed by a highly trained aircraft recognition expert in the Royal Observer Corps who happened to be on the rig at the time.

Shiva had been my first brush with the black world, the Pentagon's hidden reservoir of defence programmes – projects so secret that officially they did not exist. Since then, I'd felt the presence of other deep black projects, but only indirectly.

Looking for the black world was like looking for evidence of black holes. You couldn't see a black hole, no matter how powerful your telescope, because its pull sucked in everything around it, including the light of neighbouring stars. But astronomers knew that black holes existed because of the intense friction they generated on their edges. It was this that gave them away.

Forty years ago, the people in charge of the Air Force's hidden budget would have been quick to see the extraordinary implication of Trimble's message; that there would be no limit – no limit at all – to the potential of an anti-gravity aerospace vehicle.

The black world would have thrilled to the notion of a science that did not exist.

As I crossed from the park back into the glare of the streetlights, I knew I was quite possibly staring at a secret that had been buried more than 40 years deep.

I called Lawrence Cross, an aerospace journalist from the circuit, an ex-Jane's man, now a bureau chief for a rival publication in Australia.

Cross and I had spent long hours ruminating on the existence of US Air Force black programmes and the kind of technologies the Air Force might be pursuing in ultra-secrecy.

I liked Cross, because he had his feet squarely on the ground, was a hell of a good reporter, but wasn't your average dyed-in-the-wool-type hack. It had been a while since we had spoken.

The phone rang for ages. It was ten at night my time, eight in the morning his; and it was the weekend. I could hear the sleep in his voice when he finally picked up the handset. In the background, a baby was crying. Cross had three kids under six years old. Ninety per cent of the time he looked completely knackered.

He remained quiet as I sketched out the events of the past weeks. I told him about the article, Trimble's initial willingness to be interviewed, then the phone-call from Abelman and her insistence, on Trimble's say-so, that I drop the whole business.

'This is interesting,' he said, stifling a yawn, 'but why the long-distance call?'

'I wanted to tap you on one of your case-studies.'

'Uh huh.' He sounded wary, more alert suddenly. 'Which one?'

'Belgium,' I said. 'Wasn't there some kind of flap there a year or two back?'

'You could say that. Hundreds of people reported seeing triangular-shaped craft all over the country in two 24-hour waves – one in 1989, the other in 1990. The Belgian Air Force even scrambled F-16s to intercept them. Why the sudden interest?'

'You once told me that those craft might have been the result of some kind of secret US development effort.'

Cross laughed. 'Maybe I did. But I've had time to study the official reports since – the ones put out by the Belgian government. Those craft were totally silent. They hovered, often very close to witnesses, and they never made a sound. You may find my take on this hard to swallow, but there is no technology – no technology on earth – that could produce that kind of performance.'

'Didn't the Belgian press try to tag the sightings to Aurora?'

'Yes, but you and I know that that's crazy. Even if Aurora is real, don't tell me it can remain stationary one moment and fly Mach 7 the next. And without making a sound. Belgian radar tracked these things. The tapes show that they pulled turns of around 20 to 40 g – enough to kill a human pilot.' He paused. 'You're not seriously suggesting what I think you're suggesting, are you?'

'A 40-year US development effort, in the black, to make anti-gravity technology a reality? Why not? They were talking about it openly in 1956, Lawrence, then it dropped off the scope. Completely and utterly, like somebody orchestrated the disappearance. It makes me want to consider the possibility, at least, that someone achieved a breakthrough and the whole thing went super-classified.'

'And now you've got the bit between your teeth?'

'Something like that, yes.'

For a long moment, Cross fell silent. Then I heard him light a cigarette. In the background, I heard his wife calling him. Then he cupped the receiver, because I caught his voice, muffled, telling her he'd be there in a minute. The baby was still bawling its lungs out.

'If you break cover on this,' he said, 'you'll blow everything. For yourself, I mean.'

'Come on,' I said, 'it's a story. It may be an old story, but I'll apply the rules that I would on any other. If there's any truth in it, the answers will pop out. They usually do.'

'That's so bloody naive. If there is any truth in it, which I doubt, they'll already know you're interested and that's not going to help you one little bit. They'll stand in your way, like they may have done already with this old man . . . Trimble? If there isn't any truth in it, then you're just going to look like a fucking idiot.'

'They, Lawrence? Who's they?'

'The security people. The keepers of the secrets. The men-in-black. You know who I mean.'

I didn't. To my ears, it sounded more than a little insane.

'I'm going to bide my time,' I said, returning to the reason for the call, 'do this at my own pace. For the moment, there's no need for me to break cover. Right now, all I have to do is conduct some low-level research and keep my eyes and ears open when I'm out there in the field. No one has to know about any of this, Lawrence. All I'm asking for in the meantime is a little help. Some of your knowledge. A few facts.'

'Listen,' he said, 'there are no facts in this field; the whole business, if you want to know, is riven with disinformation, much of it, in my opinion, deliberately orchestrated. Sooner or later, you're going to have to surface and when you do, some of that crazy UFO spin is going to rub off on you. That happens and you'll never eat lunch in this great industry of ours again. Do you understand what I'm saying?'

And then his tone softened. 'I've got to go now, but if you really are hell-bent on taking this forward, you might want to try an outfit in Washington D.C. called the Integrity Research Institute. They have a handle on some of this material. Just promise me you'll keep my name out of it, okay?' And with that, he hung up.

During a trawl of the Library of Congress in the summer of 1985, Dr Paul LaViolette, a researcher badly bitten by the gravity bug, tripped over a reference in the card indexing system to a report called *Electrogravitics Systems – An Examination of Electrostatic Motion, Dynamic Counterbary and Barycentric Control.* The words immediately rang the cherries with LaViolette, because he knew they were synonyms for something that science said was impossible: anti-gravity.

LaViolette had become interested in anti-gravity during his study for a Ph.D. in systems science from Portland State University. His curiosity was aroused further when he realised that the electrogravitics report was missing from its rightful place in the library's stacks – 'like it had been lifted,' he told me later.

When LaViolette asked the librarian if she could try to locate a copy elsewhere, what began as a simple checking procedure soon developed into a full-blown search of the inter-library loan network, all to no avail.

Convinced he would never see a copy, LaViolette gave up, but weeks later the librarian called to tell him that she had managed to track one down. It had been buried in the technical library of Wright-Patterson Air Force Base in Dayton, Ohio.

There were no other copies known to exist across the whole of America.

'Whatever's in it,' she told LaViolette drily, 'it must be pretty exotic.'

When LaViolette finally got the Air Force to release the report, exotic seemed about right. *Electrogravitics Systems*, drafted in February 1956, the same year Trimble and his colleagues began pronouncing publicly on anti-gravity, contained details of what appeared to be a Mach 3 anti-gravity aerospace vehicle designed as a fighter-interceptor for the US Air Force.

The fact that the report had been located at the Air Force's premier research and development facility instantly made *Electrogravitics Systems* an intriguing piece of work. That the USAF also appeared to have sat on it for almost 40 years only served to underscore its importance, LaViolette and his associates felt.

I had managed to track LaViolette through the Integrity Research Institute. The institute sat on the edges of a sub-culture of researchers who relentlessly picked over the anti-gravity issue. Type 'anti-gravity' into a Net search-engine and an outpouring of conspiracy-based nonsense on the government suppression of anti-gravity technology usually popped up on-screen. In the sober world of defence journalism, I'd never remotely envisaged a day when I'd have to involve myself in such matters.

Now, recalling Cross' warnings about professional suicide, I was reluctant to take the plunge, but realised that I had no choice.

I spoke to a colleague of LaViolette's called Tom Valone who mercifully never pressed me on the reasons why I was so interested in the anti-gravity business. A few days later, a copy of *Electrogravitics Systems* duly arrived through the post.

Electrogravitics Systems had not been prepared by the Air Force, it turned out, but by an outfit called the Gravity Research Group, Special Weapons Study Unit, a sub-division of Aviation Studies (International) Limited, a British organisation.

I did some checking. Aviation Studies had been a think-tank run by Richard 'Dicky' Worcester and John Longhurst, two talented young aerospace analysts who in the mid-1950s produced a cyclo-styled newsletter that had a reputation for close-to-the-knuckle, behind-the-scenes reporting on the aerospace and defence industry.

From the start, the *Electrogravitics Systems* report adopted a lofty tone: 'Aircraft design is still fundamentally as the Wrights adumbrated it, with wings, body, tails, moving or flapping controls, landing gear and so forth. The Wright biplane was a powered glider, and all subsequent aircraft, including supersonic jets of the 1950s are also powered gliders.'

The writers went on to say that insufficient attention had been paid to gravity research and that the 'rewards of success are too far-reaching to be overlooked.'

But while it quickly became clear that the report was not authored by the USAF, or apparently even commissioned by it, it did appear to be remarkably well informed about the progress of gravity – and anti-gravity – research in the United States and elsewhere; something that LaViolette and Valone felt might have accounted for its burial for so long in the Wright-Patterson technical library.

The words 'anti-gravity' were not used in the report, presumably because this pop-scientific term was felt to be too lurid for the report's paying subscribers. To those less fussed, however, the term 'electro-gravitics' seemed pretty much synonymous with anti-gravity, even by Aviation Studies' own definition:

'Electrogravitics might be described as a synthesis of electrostatic energy used for propulsion – either vertical propulsion or horizontal or both – and gravitics, or dynamic counterbary, in which energy is also used to set up a local gravitational force independent of that of the earth.'

Counterbary is subsequently defined as 'the action of levitation where gravity's force is more than overcome by electrostatic or other pro-pulsion.'

Anti-gravity, then.

The report echoed what had already been made clear in the Gladych and *Interavia* articles: that anti-gravity was of real interest to just about every US aerospace company of the day.

'Douglas has now stated that it has counterbary on its work agenda, but does not expect results yet awhile. Hiller has referred to new forms of flying platform. Glenn Martin say gravity control could be achieved in six years, but they add it would entail a Manhattan Project type of effort to bring it about. Sikorsky, one of the pioneers, more or less agrees with the Douglas verdict . . . Clarke Electronics state they have a rig, and add that in their view the source of gravity's force will be understood sooner than some people think. General Electric is working on electronic rigs designed to make adjustments to gravity – this line of attack has the advantage of using rigs already in existence for other defence work. Bell also has an experimental rig intended, as the company puts it, to cancel out gravity, and Lawrence Bell has said that he is convinced practical hardware will emerge from current programs. Grover Leoning is certain that what he referred to as an electro-magnetic contra-gravity mechanism will be developed for practical use. Convair is extensively committed to the work with several rigs. Lear Inc, autopilot and electronic engineers, have a division of the company working on gravity research and so also has the Sperry division of Sperry-Rand. This list embraces most of the US aircraft industry. The remainder, Curtiss-Wright, Lockheed, Boeing and North American have not yet declared themselves, but all these four are known to be in various stages of study with and without rigs.'

Everybody, in other words, had had a finger in the pie.

But where, I asked, was the evidence 40 years on? It wasn't simply that the results of all this activity had failed to yield hardware; no corporate history of any these companies that I'd ever read in all my years as a journalist in the industry even hinted that gravity, anti-gravity, electro-gravitics, counterbary – call it what you will – had ever been of interest to any of them.

And then, on page three of the *Electrogravitics Systems* report I saw the mention of Thomas Townsend Brown – the 'Townsend T. Brown' of the *Interavia* piece in the Jane's library.

The report referred to Brown's 'Project Winterhaven'. Apparently, Brown had invented a whole new approach to the mechanics of flight.

This notion of electrogravitic lift supposedly worked on the principle that a plate-like object charged positively on one side and negatively on the other would always exhibit thrust towards the positive pole, i.e. from negative to positive. If the plate is mounted horizontally and the positive pole is uppermost, the object will in effect lose weight, because it will want to rise skyward.

I was in no rush to judge the validity of the physics. But I'd worked in the business long enough to know that the military would be quick to dismiss anything that wasn't practical. So the key question for me was how the military had viewed Brown's work.

I called the Integrity Research Institute and asked Tom Valone to send me as much information as his organisation possessed on Brown.

Then, over the next few weeks, my life assumed a pattern. In the dead hours, long after the phone stopped ringing, I'd head down into the basement, fire up the computer and hit the Net, staying on-line deep into the night.

My great regret was that I couldn't contact George S. Trimble directly. Had I done so, I knew that Abelman would have gone ballistic. She'd told me to stay away from him and she had the power to ensure that I became an outcast if I didn't. Lockheed Martin was a large company and its word would spread quickly.

In that respect, Cross was right. Pretty soon, no one would want to talk to a technology hack who was running around asking madcap questions about anti-gravity.

Better, then, to do what I was doing; keep a lid on it. Here, in the silence of the night, I could roam the Internet and remain anonymous. Besides, it really did seem to be a case of reviewing the evidence and following up the clues – clues that had apparently never registered with the experts who had been reporting on all this for years.

That in itself was seductive. Little by little, I could feel myself being pulled in.

Chapter 3

Thomas Townsend Brown was born into a prominent family from Zanesville, Ohio, in 1905, two years after the Wright Brothers, residents of nearby Dayton, propelled their 'Flyer', the first aircraft capable of sustained powered flight, into the sky at Kitty Hawk, North Carolina. From an early age, Brown exhibited traits that would later come to mark his work as a scientist. As a child, he built a workshop in his back yard and at the age of ten was already receiving signals from across the Atlantic on a home-made radio-receiver. At 16, he was broadcasting from his own radio station.

For all his prowess as an inventor, Brown appears to have been a somewhat recalcitrant student. In 1922, he was enrolled at the California Institute of Technology (Cal Tech), but soon fell into a disagreement with his teachers over the time allowed by the Institution for laboratory work, something he lived for. After failing his first year exams in chemistry and physics, he persuaded his father to sponsor the construction of a laboratory of his own. With no expense spared, a lab was installed on the second floor of their new house in Pasadena.

The quid pro quo, apparently, was that 'T.T.' had to receive home tuition to boost his grades. Brown subsequently reported that he made considerable progress in chemistry, while still finding time to devise an X-ray spectrometer for astronomical measurements. It was at this point that he began to develop his first theories about electrogravitation. In essence, Brown believed that gravitation might be a form of radiation, much like light, with a 'push' as opposed to a 'pull' effect.

'Word of this got out among my classmates,' he related in a document called 'A Short Autobiography', which I found on the Net, 'and although shunned and made fun of by the professors, I was nevertheless called to the attention of Dr Robert Andrew Millikan, Director of Cal Tech and, incidentally, my first physics teacher, who explained to me in great detail, why such an explanation of gravitation was utterly impossible and not to be considered.'

In 1923, Brown transferred to Kenyon College, Gambier, Ohio, and the following year switched to Ohio's Denison University where he came under the tutelage of physicist and astronomer Dr Paul Alfred Biefeld, a former classmate of Einstein's in Switzerland. In earlier experimentation, Brown had made the startling discovery that a Coolidge X-ray tube exhibited thrust when charged to high voltage. It took Brown a while to realise that the motion was not caused by the X-rays themselves, but by the electricity coursing through the tube. Brown went on to develop a device he called the 'Gravitor', an electrical condenser sealed in a Bakelite case, that would exhibit a one per cent weight gain or a one per cent weight loss when connected to a 100-kilovolt power supply.

In 1929, Brown wrote up his discoveries in a paper entitled *How I Control Gravitation* and speculated as to how his invention might one day be used:

'The Gravitor, in all reality, is a very efficient motor. Unlike other forms of motor, it does not in any way involve the principles of electro-magnetism, but instead it utilises the principles of electro-gravitation.

'A simple gravitor has no moving parts, but is apparently capable of moving itself from within itself. It is highly efficient for the reason that it uses no gears, shafts, propellers, or wheels in creating its motive power. It has no internal mechanical resistance and no observable rise in temperature. Contrary to the common belief that gravitational motors must necessarily be vertical acting, the gravitor, it is found, acts equally well in every conceivable direction.'

This ability to manipulate the force in all axes opened up the Gravitor's potential to aviation. Brown's research led him to the development of a shape that was the most efficient in the production of electrogravitational lift: that of a perfect disc or saucer.

The way he proposed to control the craft was by dividing the disc into segments, each of which could be selectively charged. By moving the charge around the rim of the saucer, it would, Brown said, be possible to make it move in any direction.

I had to remind myself that this was the 1920s.

The aviation industry was still absorbing the technical lessons of the First World War, the biplane was king and fighters, typified by the Boeing Model 15A, a leading US design of the mid-1920s, were struggling to attain top speeds of 160 mph.

But although Brown's principles seemed to have been confirmed by observation – something, after all, was causing his condenser plates to move, however much the nay-sayers denigrated the possibility that it was due to an electrogravitic reaction – he had no way of generating and maintaining an electrical charge that would keep a small model craft in the air, let alone to manoeuvre it. This force was calculated to be in the region of 50 kilovolts, 200 times more than the charge that came from my wall-socket.

Yet these were early days. What was astounding was that Brown had developed a concept for an air vehicle, shaped in the form of a disc, years before anyone had coined the term 'flying saucer'.

This, I thought, had to be more than happenstance, especially as the military were showing signs that they considered Brown to be not just a visionary, but a practical man of invention; someone they could turn to to develop nuts-and-bolts hardware.

In September 1930, Brown joined the US Navy Reserve, but instead of going to sea was given orders to report to the Naval Research Laboratory (NRL) in Washington D.C. In due course, he transferred to the NRL's 'Heat and Light Division' and here he carried on the experiments that he started in Ohio.

Experiments were conducted which seemed to prove the concept of gravitation which he had first postulated at Cal Tech in 1923.

In 1932, Brown served as staff physicist on the Navy's Gravity Expedition to the West Indies. But by 1933, the effects of the Great Depression were forcing cutbacks within the Navy and Brown, suddenly finding himself out of a job at the NRL, was driven to seek work in industry, which he interleaved with his duties as a reserve naval officer.

With war clouds gathering in Europe, Brown was drafted full-time into the Navy and assigned to acoustic and magnetic minesweeping research, a field for which he seemed to have a great natural aptitude. He developed a method for maintaining the buoyancy of minesweeping cables, designed to trigger magnetic mines with a strong electrical force field, whilst simultaneously shielding them from blast effects. Brown took out a patent on the idea, which was immediately classified.

By 1940, he was appointed head of the Navy Bureau of Ships' mine-sweeping research and development activity and it was during this period that he experimented with 'degaussing' – a method for cancelling a ship's magnetic field. This was a critical breakthrough as it was a ship's magnetic field that triggered a new breed of German mines, weapons that the Nazis had developed in quantity.

There was no question that Brown was by now well plugged into the military's research and development community, and that his opinions were given great credit.

It is ironic, then, that it is at this point that his credibility is most open to question, thanks to his supposed association with the so-called 'Philadelphia Experiment'.

In this, according to legend, the US Navy spirited one of its warships, complete with its crew complement, into another dimension.

The Philadelphia Experiment was made famous in a book of the same name by Charles Berlitz and William Moore, writers who have forged names for themselves with tales of paranormal mysteries. Even they confessed that it was 'questionable whether Brown was really ever very heavily involved in the Philadelphia Experiment project' – a comment that assumes this incident took place at all.

But the association is ingrained. Run a check with a search-engine and Brown is up there, his name highlighted, nine times out of ten, right alongside it.

I tried to manoeuvre around this obstacle, this blip in Brown's otherwise blemishless wartime record, but whichever way I turned, it wouldn't go away.

Its effect was powerful and immediate. Up until this moment, I'd trawled cyberspace and decades-old documents in the growing conviction that this unusually gifted engineer was an important, long-overlooked link in the anti-gravity mystery highlighted by the Gladych and *Interavia* articles.

But the instant the Philadelphia Experiment started to be mentioned in the same breath as Brown, it made me want to drop him like a stone.

It was just as Cross had said it would be. I could feel the association, with its whiff of paranoia and conspiracy, threatening to act like a contaminant on my thinking.

Better to avoid it altogether.

Everything had been going so well. Brown was university-educated, he'd worked in the aviation industry, he'd been embraced by the military and he'd done classified work. And then, the Philadelphia Experiment had come along.

Shit.

The substance of the story is that while experimenting with techniques in 1943 to make navy ships invisible to radar through the use of intense electromagnetic fields, a destroyer, the USS *Eldridge*, disappeared from its berth in Philadelphia and reappeared moments later at a Navy yard in Norfolk, Virginia, 250 miles away. During this time-lapsed interval, the crew were supposedly transported into a 'parallel dimension', an experience that drove many of them insane.

Brown's credentials at the time, coupled with his refusal in the years since altogether to dismiss the Philadelphia Experiment as hokum, have helped to stoke the myth. And so it has stuck to him like glue.

In the silence of the basement, confronted by this dead-end, I began to wonder about that.

If the original authorities on the story, Berlitz and Moore, had had their own doubts about Brown's involvement in the Philadelphia Experiment, how had he come to be linked so inextricably to it?

And then I thought of something else; something Cross had said. That this whole field was riven with disinformation, some of it, in his opinion, deliberately managed. The Soviets had even coined a term for it: *disinformatsiya*.

During World War Two and the long decades of the Cold War the Russian military had used *disinformatsiya* to achieve key tactical and strategic objectives.

Far from avoiding the Philadelphia Experiment, which was what my every professional instinct yelled I should be doing, I came to the uncomfortable conclusion that I ought to do precisely the opposite.

Tentatively, I began to surf the Net for obscure websites that told the story.

There are two versions of what the Philadelphia Experiment was trying to achieve. One holds that the Navy was testing a method that would make ships invisible to radar, the other was that it was attempting to develop some kind of *optical* cloaking device as well; that by generating an intense electromagnetic field around the *Eldridge*, it would distort both light and radar waves in its vicinity, rendering the ship invisible to both sensor systems and the human eye.

To do this, the legend stated, the *Eldridge* was equipped with tons of electronic equipment. These, the story had it, included massive electrical generators, several powerful radio frequency transmitters, thousands of power amplifier tubes, cabling to distribute the energy around the ship and special circuitry to tune and modulate the fields.

During the first test, which is supposed to have taken place in July 1943, the *Eldridge* is said to have become invisible at its berth, the only sign that it was there being a trough of displaced water beneath the hull. Though classified a success, the ship remaining invisible for around 15 minutes, the crew experienced side-effects, including nausea, disorientation and memory loss.

If we can for the moment assume that there's some truth in the story this would have been quite unsurprising to those in charge of the science, since the nerve impulses by which the brain operates are signalled electrically. In effect, just as earlier iterations of the equipment had demonstrated on enemy mines, the field-generators would have 'degaussed' the brains of anyone within the area of influence, causing them temporary and, in some cases, permanent damage.

In such circumstances, I could conceive of reasons why the Navy might want to plant misleading stories about the *Eldridge* – the more so, since it involved stealth.

It remains highly sensitive technology, but the use of electromagnetic fields in generating radar-stealth is a technique well known to military science today.

But during World War Two – if Brown had discovered a means of shielding US Navy ships from enemy radar (or even for marginally reducing their radar signature) – it would have rung right off the classification scale.

The last thing the Navy would have wanted was sailors running around complaining of headaches and revealing how they had come by them.

The next experiment supposedly went even further. In October 1943, after the Navy had replaced the first crew, a second test on the *Eldridge* was carried out.

This time, soon after the generators were switched on, the ship shimmered for a few seconds, remaining visible in outline only. Then there was a blinding flash and it vanished altogether. The legend states that it was transported briefly to a berth in Norfolk, Virginia, before making its way back to the yard at Philadelphia.

When investigators went on board, they found that some of the crew had been swallowed by the aether that had momentarily consumed the *Eldridge*. These men were never seen again.

Those that did make it back were either made dangerously ill by their adventure, experiencing intense nausea and headaches for years, or were driven mad by it.

Weirdest of all, five of the crew were found fused into the metallic structure of the ship, which had materially transmuted to accommodate them.

These were the broad 'facts' as presented to Berlitz and Moore, who were first alerted to the mystery by a man who claimed to have been on the *Eldridge*. They were supported by a dangerously eccentric source identified variously as Carlos Allende and Carl Allen, who claimed to have witnessed the whole thing from the deck of the SS *Andrew Furuseth*, a merchant marine ship berthed close to the *Eldridge* in the Philadelphia navy yard.

This was in the mid-1970s, more than 30 years after the purported events. When the book came to be written, Brown became an inextricable part of the myth.

The only other thing I knew about disinformation, certainly as it had been practised by the Soviets, was that it worked best when mixed in with a little truth.

Intuitively, therefore, I felt Brown must have been somewhere in the vicinity of the 'experiment' when it was supposed to have happened. Beyond that, I could draw no other conclusions, so I returned to the documented facts of his career.

In 1942, he was appointed head of the Atlantic Fleet Radar Materiel School and Gyrocompass School in Norfolk, Virginia, a position that would have made him privy to some of the most highly classified technical secrets of the day. Whatever work he was engaged on there, it appears to have taken its toll, since the following year he suffered a nervous breakdown and was discharged from the Navy.

There is something neatly synergistic in this unfortunate development, which appears to have been real enough, with the mystery surrounding the USS *Eldridge*. One suggestion is that whatever work Brown was engaged in at the time, be it in the radar or minesweeping field, it caused him temporary memory loss.

Whilst it stretched credulity to believe in stories of optical invisibility, tele-transportation and parallel dimensions, I did find it possible to envisage a scenario in which the legend of the *Eldridge* had grown out of the secrecy of Brown's legitimate work in the radar field; in much the same way that the supposed ability of RAF night fighter pilots to see in the dark by eating carrots stemmed from a childishly simple British ruse to protect its radar secrets at around the same time.

Despite Brown's illness – and whatever really happened in the Philadelphia and Norfolk shipyards – it appears to have had no long-term detrimental effect on his standing in the eyes of the Navy. In 1944, he went back to work as a radar consultant in Burbank, California, at Lockheed's Vega Division, which was responsible for development of the Navy's PV-2 Harpoon and P-2V Neptune anti-submarine patrol bombers.

At war's end, Brown moved to Pearl Harbor, Hawaii, where he once again resumed the anti-gravity work that had driven his research efforts as a young man.

It seemed to me from this period – long before the story of the Philadephia Experiment emerged to complicate the picture – as if two portraits of Brown had been painted and were now in circulation: one portraying him as a mildly eccentric inventor with some hare-brained ideas about negating the forces of gravity; the other showing him to be a man responsible for some of the most highly classified research of the war.

Seen through this ambivalent prism, the story of the Philadelphia Experiment has helped to perform a very important function.

By 1980, it had managed to tip Brown over the edge; make him a wholly discredited figure in the eyes of science.

That left me with the uncomfortable feeling that the story had been carefully stage-managed. If so, why? And why so long after the supposed events had taken place?

The next occasion Brown is reputed to have demonstrated his Gravitor and tethered flying discs was to Admiral Arthur W. Radford, Commander-in-chief of the US Pacific Fleet, in Hawaii in 1945. Brown had by now left California and was resident in Hawaii, where he worked temporarily as a consultant at the Pearl Harbor Navy Yard.

His demonstrations supposedly failed to impress the Navy, but there is no known official record of its reaction. One account is that it 'refused funding for further research because of the negative opinion of other scientists.' Another relates the story of his room being broken into after the Pearl Harbor demonstration and the theft of his notebooks. According to this variant, which was related by a close friend of Brown's, Josh Reynolds, the Navy returned the books days later, adding that it remained uninterested. The reason given was that the 'effect' was not due to electrogravitation, but to 'ionic wind' – this, despite Brown's view that he had conclusively proven otherwise with experiments under oil back in the 1920s.

From 1945 until 1952, little is known about Brown's activities, but based on what happened next, it is clear that along the way he returned from Hawaii to Los Angeles, where he established the Townsend Brown Foundation.

In 1952, the Foundation received an unannounced visit by Major General Victor E. Bertrandias of the US Air Force. I learned of this thanks to a recovered transcript of a phone conversation between Bertrandias and another USAF general called Craig, whose job description was not immediately clear.

The transcript had been forwarded to me by Tom Valone. It made clear that Bertrandias was astounded by what he witnessed.

'It sounds terribly screwy, but Friday I went down with Lehr, a man named Lehr, to a place called the Townsend Brown Foundation, and believe it or not, I saw a model of a flying saucer,' Bertrandias reported.

'No,' Craig replied, apparently without irony.

'I thought I should report it,' Bertrandias told him. 'There was a lot of objection to my getting in there by the party that took me and the thing frightened me – frightened me for the fact that it is being held or conducted by a private group. I was in there from about one-thirty until about five in the afternoon and I saw these two models that fly and the thing has such a terrific impact that I thought we ought to find out something about it – who these people are and whether the thing is legitimate. If it ever gets away I say it is in the stage in which the atomic development was in the early days.'

'I see,' General Craig replied. The man's urbane delivery earmarked him, to me at least, as someone big in Air Force intelligence.

'I was told that I was not to say anything about it,' Bertrandias burbled on, 'but I'm afraid that all that I heard made me believe in it and these were not school boys. It was conducted in rather an elaborate office in Los Angeles. I thought I should report it to you.'

'Yes, well I'll look into it and see what I can find out,' Craig told him, shortly before hanging up.

It seems that what Bertrandias had stumbled upon, a little before the Townsend Brown Foundation was ready to pitch it formally to the military, was Project Winterhaven – the distillation of all Brown's ideas into a blueprint for a manned anti-gravity fighter, built in the shape of a disc and capable of Mach 3, twice the speed of the leading jet-powered interceptor of the day.

Years later, LaViolette had found extensive references to Winterhaven in the sole remaining copy of the *Electrogravitic Systems* report the Congressional librarian had tracked down to the technical library at Wright-Patterson air force base.

I was rapidly coming full circle.

I needed to see the report on project Winterhaven, but remembering the difficulty LaViolette had had in tracking down the one remaining copy of *Electrogravitics Systems*, I couldn't begin to think how or where I would lay my hands on one. I imagined, given the Air Force's evident excitement about Brown's work, that if there were any copies of *Winterhaven* left, they were buried somewhere deep.

I rang Valone in Washington – my third call in as many weeks – and this time got the question I'd been dreading. Was *Jane's* running some kind of investigation into anti-gravity?

I told him it wasn't; that my interest in Brown's work was born of personal curiosity, nothing more, and that I'd appreciate it if he'd keep the matter to himself.

Valone seemed quite unfazed by this request. In the murky world of anti-gravity research, I realised it was probably par for the course.

The conversation drifted inevitably towards Townsend Brown and *Winterhaven*. It was then that I mentioned how I'd give my eye-teeth to know what was in it.

A week later, to my absolute astonishment, a copy of *Winterhaven* (registered copy No. 36) landed on my desk. Attached to the front page was a note from Valone. Thanks to the Freedom of Information Act, there was no need for me and my teeth to part company, he said. I smiled. The man was a one-stop shop.

He assured me, too, that no one else bar him and LaViolette would ever know that I was interested in the subject-matter.

Nagged by the ease with which the military had relinquished its grip on *Winterhaven*, I started reading.

In his pitch to the military, on page six, Brown had written: 'The technical development of the electrogravitic reaction would usher in a new age of speed and power and of revolutionary new methods of transportation and communication.

'Theoretical considerations would predict that, because of the sustained acceleration, top limits of speed may be raised far beyond those of jet propulsion or rocket drive, with possibilities of eventually approaching the speed of light in "free space". The motor which may be forthcoming will be essentially soundless, vibrationless and heatless.'

I made notes. This was language that was similar in tone to that expressed by George S. Trimble and his quoted contemporaries in the 1956 Gladych article. Trimble, who 40 years later had cancelled my interview with him, because he appeared – to Abelman's mind, at least – to be too afraid to talk.

The attributes of the technology were also bang in line with the characteristics exhibited by the flying triangles observed over Belgium.

The little model discs that had so impressed General Bertrandias 'contain no moving parts and do not necessarily rotate while in flight,' the Winterhaven report stated. 'In atmospheric air they emit a bluish-red electric coronal glow and a faint hissing sound.'

Project Winterhaven, then, offered a systematic approach for the establishment of a US anti-gravity programme – echoing the origins of the US atomic bomb project a decade earlier and the path that Trimble had advocated in 1956.

In the report, Brown recommended starting modestly with 2 ft discs charged at 50 kilovolts, then proceeding to 4 ft discs at 150 kV and finally to a 10 ft disc at 500 kV, with careful measurements being made along the way of the thrust generated.

This was a sensible, structured approach. It showed that Winterhaven was ready to transition from the drawing board to the technology demonstration phase.

Full-blown development wouldn't be far behind.

Brown had also formulated a method for generating the required high voltages for a free-flying disc – the problem that had earlier stymied its practical development into a manned aircraft – by means of a 'flame-jet generator'. This was a jet engine modified to act as an electrostatic generator capable of providing up to 15 million volts to the skin of the craft. The *Interavia* article, I remembered, had talked of trials involving 3 ft discs run in a 50 ft diameter air course under a charge of 150 kV 'with results so impressive as to be highly classified.'

Brown's flying saucer now had a power-source.

In one way and another things seemed to be gaining momentum.

But it was short-lived.

Days after General Bertrandias' conversation with his spook colleague General Craig, the documents trail showed that the Air Force Office of Special Investigations, AFOSI, was well and truly onto the Brown case. AFOSI was the Air Force equivalent of the FBI – an indication that Craig had taken Bertrandias' interpretation of his visit to the Brown Foundation seriously. Once again, it was Valone who furnished me with the relevant papers.

They showed that AFOSI uncovered a copy of a report prepared by one Willoughby M. Cady of the Office of Naval Research, the ONR, in Pasadena, California, entitled: 'An Investigation Relative to the Townsend Brown Foundation'.

The report, which was forwarded by AFOSI to Major General Joseph F. Carroll, Deputy Inspector General of the USAF, made disappointing reading all round.

'Mr Brown claims that a gravitational anomaly exists in the neighbourhood of a charged condenser,' Cady wrote in his conclusions, adding: 'This effect has not been well documented by Mr Brown.'

Cady went on to say that there was no electrogravitational link, merely a disturbance of the air around the model saucers stimulated by the electrical charge and it was this that was causing them to be propelled upwards and forwards.

Any hope that this technology might be of use in aerospace propulsion was damned by Cady. 'If the efficiency of conversion of fuel energy into electrical energy is 21 per cent, the overall efficiency of propulsion of the model flying saucers by the electric wind is 0.3 per cent. This compares with about 25 per cent for a propeller-driven airplane and 15 per cent for a jet airplane.'

If Cady's conclusions were clear to me, they would have been crystal to the Navy and the Air Force. Brown's flying saucer would have had trouble getting off the ground, let alone cruising at Mach 3. The whole 'science' of electrogravitics, Cady was telling his superiors, was a waste of time, effort and money.

The Navy and its old foe the Air Force would have concluded that they were both far better off sticking with jets.

Even under the auspices of a deeply classified programme, an arrangement by which one branch of the armed forces might be unaware of projects taking place in another, it was hard to see Brown's work surviving such an appraisal.

On 22 September 1952, six months after General Bertrandias' telephone call to General Craig, documents showed that the Air Force downgraded its interest in the work of the Townsend Brown Foundation from 'confidential' to 'unclassified'.

This explained why the Pentagon's archives had contained a declassified copy of *Winterhaven*.

Despite the view of LaViolette, Valone and many others that this was the moment Project Winterhaven went super-classified, forming the basis of a number of 'black programmes' in the anti-gravity field from the mid-1950s onwards, I found this hard to believe. Although LaViolette's theory dovetailed neatly with the ringing silence that followed the zealous rhetoric of George Trimble and his colleagues in the US aerospace industry on the subject of anti-gravity in 1956, there was no evidence at all to support it.

Nor, crucially, did LaViolette's thesis explain Brown's behaviour from this period onwards. For although it was obvious that he knew how to keep a secret – his wartime work would have instilled in him the need for the rigours of secrecy – if the ONR report was merely a blind, his subsequent actions went way beyond attempts to lead investigators away from a buried programme within the US Navy or Air Force or both. The energy with which he now set out to prove his electrogravitational theory seemed, in fact, to stem specifically from the comprehensive rejection of his experiments by the US military. For the rest of his life, Brown, who was universally described as quiet, unassuming, honest and likeable, behaved like a man driven to prove his point.

In 1955, he went to work for the French aerospace company SNCASO – Société Nationale de Constructions Aéronautiques du Sud-Ouest.

During this one-year research period, he ran his discs in a vacuum. If anything, they worked better in a vacuum, something that prompted SNCASO, which was interested in exploiting Brown's work for possible space applications, to offer him an extension of his contract. But in 1956, SNCASO merged with its counterpart Sud-Est and its new bosses saw little future in the space business. They wanted to build aeroplanes; real ones, with wings and jet engines.

His contract terminated, Brown returned to America, where he helped found the National Investigations Committee on Aerial Phenomena (NICAP), an unofficial study group set up to analyse the growing body of UFO sightings across the American continent and elsewhere. It was Brown's belief that the study of UFOs, many of which exhibited characteristics similar to his discs, could shed light on their propulsion methodology and this, in turn, could be exploited for science and space travel. The trouble was, of course, that this alienated him even more from the mainstream.

It also further alienated him from me. As I'd immersed myself in Brown's life and work, I'd wondered more and more if his 'discovery' of electrogravitics had had something to do with the emergence of the whole UFO phenomenon in the 40s, if these 'alien' craft had in fact been top-secret aerospace vehicles propelled by a power-source that science even today refused to recognise. This, however, did not gel with a man who went on to found a UFO study group.

In 1957, Brown was hired as a consultant to continue his anti-gravity work for the Bahnson Company of North Carolina and in 1959 he found himself consulting for the aerospace propulsion giant General Electric.

I found little corroborating evidence for Brown's activities during these years.

When he went into semi-retirement in the mid-1960s, Valone and LaViolette saw this as a signal that, in effect, he had been been bought off by the military, especially as he hardly touched electrogravitics again. His last great interest involved a series of ultimately successful attempts to draw stored electrical energy – albeit in minute quantities – from common or garden rocks . . .

Unquestionably a highly gifted and unusual man, Brown died in relative obscurity in 1983.

In trying to draw lessons from his story, lessons that might perhaps help explain the sudden outpouring of interest in anti-gravity by America's leading aerospace designers in the late 1950s, I found myself trying to put the pieces of the puzzle together. The problem was, however I went about it, they wouldn't form into any recognisable picture.

If the Navy had developed Brown's experiments into a fully fledged anti-gravity programme, why had its premier research arm, the ONR, gone out of its way to dismiss them? And if this was an elaborate ruse to throw others off the scent – including its old rival the US Air Force – why did Brown continue in his work, eventually taking it to another country? If electrogravitics was classified, this would have handed it on a plate to another nation, which was madness.

There was, conceivably, a chance that Brown's work had been hijacked by the US military without his knowledge at the end of World War Two, but this too seemed unlikely. Given Brown's wartime clearances, it would have been simpler to have sworn him to secrecy after the Hawaii experiments, had the Navy been that impressed with them.

There was another, remote possibility – that Brown's work had been rejected by the military, not because it was hokey or mad, but because its principles were already known to them – and, perhaps, therefore already the subject of advanced development activity. Had this been the case, it might well have explained why, a few years later, Bell, Convair, Martin and so many other companies – equally ignorant of this activity – aired their views on anti-gravity unchecked for a number of months, as recorded in the *Interavia* article, until someone, somewhere ordered them to silence.

As I kicked this notion around, I found sympathy for it from an unlikely source.

In the appendix section of a compendium of UFO sightings, I found a secret memorandum dated 23 September 1947 from Lieutenant General Nathan Twining, head of the US Army Air Forces' Air Material Command (AMC), to Brigadier General George Schulgen, a senior USAAF staff officer in Washington.

The memorandum had been declassified and released into the public domain only in the late 1970s. I'd found it during another late-night trawl of my burgeoning archive, a cup of strong coffee in my hand to keep me awake.

In the memo, Twining states in his 'considered opinion' that the rash of UFO sightings in America during the summer of 1947 – the first real wave, as it turned out – had been 'something real and not visionary or fictitious'.

He continued: 'There are objects . . . approximating the shape of a disc, of such appreciable size as to appear to be as large as a man-made aircraft.' These discs exhibited a number of common characteristics, amongst them a metallic or light-reflecting surface, an absence of any jet-trail and no propulsion sound.

In addition, they were capable of 'extreme rates of climb, maneuverability (particularly in roll), and action which must be considered evasive when sighted.'

For an Air Force officer – a senior one, even – to admit to the reality of UFOs was quite an admission, but it was not unique. Other then-classified memos from other USAAF officers of the day show them to have been similarly perplexed by the disc sightings phenomenon across America from mid-1947 onwards.

UFO researchers had seized on these documents as evidence that the USAAF (and from 1947 its successor, the USAF) recognised the reality of UFOs while officially pooh-poohing them – a blanket denial that remains policy to this day.

My attention, however, had been drawn to something quite else.

As an aerospace analyst, and in light of the Brown research, I found the key part of Twining's memo in what followed. UFO researchers had concentrated entirely on the main thrust of the memo: Twining's acknowledgment of the reality of UFOs.

The part that interested me, the postscript, added almost as an afterthought, had been totally overlooked. It so absorbed me that I never noticed the pencil on which I set down my cup of lukewarm coffee. It tipped over and the drink spilled, spreading across the open pages like ink on a blotter. I swore, mopped it up as best I could and returned to the text. The stain, as poor luck would have it, had covered the paragraph I was interested in, making the task of rereading it a painfully slow affair. I uttered the words aloud, enunciating them slowly so I knew I had read it right.

'It is possible,' Twining wrote, 'within the present US knowledge – provided extensive detailed development is undertaken – to construct a piloted aircraft which has the general description of the object above which would be capable of an approximate range of 7000 miles at subsonic speeds.'

As head of Air Materiel Command, the branch of the USAAF that assumed responsibility for all USAAF aircraft development, Twining – who would have written the memo under the best advice from his subordinates – should have known what he was talking about. The problem was that 50 years this side of Twining's memo, at the end of the 20th century, I still knew of no combat aircraft capable of the performance characteristics Twining had elucidated. Even the Lockheed

---

c. There is a possibility that some of the incidents may be caused by natural phenomena, such as meteors.

d. The reported operating characteristics such as extreme rates of climb, maneuverability (particularly in roll), and action which must be considered evasive when sighted or contacted by friendly aircraft and radar, lend belief to the possibility that some of the objects are controlled either manually, automatically or remotely.

e. The apparent common description of the objects is as follows:

(1) Metallic or light reflecting surface.

Basic Ltr fr CG, AMC, WF to CG, AAF, Wash. D. C. subj "AMC Opinion Concerning "Flying Discs".

(2) Absence of trail, except in a few instances when the object apparently was operating under high performance conditions.

(3) Circular or elliptical in shape, flat on bottom and domed on top.

(4) Several reports of well kept formation flights varying from three to nine objects.

(5) Normally no associated sound, except in three instances a substantial rumbling roar was noted.

(6) Level flight speeds normally above 300 knots are estimated.

f. It is possible within the present U. S. knowledge — provided extensive detailed development is undertaken — to construct a piloted aircraft which has the general description of the object in subparagraph (e) above which would be capable of an approximate range of 7000 miles at subsonic speeds.

g. Any development in this country along the lines indicated would be extremely expensive, time consuming and at the considerable expense of current projects and therefore, if directed, should be set up independently of existing projects.

h. Due consideration must be given the following:

(1) The possibility that these objects are of domestic origin - the product of some high security project not known to AC/AS-2 or this Command.

(2) The lack of physical evidence in the shape of crash recovered exhibits which would undeniably prove the existence of these objects.

HEADQUARTERS
AIR MATERIEL COMMAND

TSDIN/BGM/1g/6-4100
Wright Field, Dayton, Ohio

SEP 2 3 1947

SUBJECT: AMC Opinion Concerning "Flying Discs"

TO: Commanding General
Army Air Forces
Washington 25, D. C.
ATTENTION: Brig. General George Schulgen
AC/AS-2

1. As requested by AC/AS-2 there is presented below the considered opinion of this Command concerning the so-called "Flying Discs". This opinion is based on interrogation report data furnished by AC/AS-2 and preliminary studies by personnel of T-2 and Aircraft Laboratory, Engineering Division T-3. This opinion was arrived at in a conference between personnel from the Air Institute of Technology, Intelligence T-2, Office, Chief of Engineering Division, and the Aircraft, Power Plant and Propeller Laboratories of Engineering Division T-3.

2. It is the opinion that:

a. The phenomenon reported is something real and not visionary or fictitious.

b. There are objects probably approximating the shape of a disc, of such appreciable size as to appear to be as large as man-made aircraft.

U-2 spyplane (which first flew in 1955, but is still in service), a single-engined aircraft intended for ultra-long endurance missions (and hardly, therefore, a manoeuvring type like the discs reported to Twining), had a maximum unrefuelled range in excess of 3,000 miles – 3,500, perhaps, on a good day.

And, needless to say, no one had ever owned up to building a flying saucer.

To construct the 'aircraft' that Twining had in mind, he told Gen Schulgen: 'Any developments in this country along the lines indicated would be extremely expensive, time-consuming and at the considerable expense of other projects and therefore, if directed, should be set up independently of existing projects.'

Placing such a project outside the mainstream of aerospace science brought to mind the effort that had resulted in the development of the first atomic bomb. In 1956, George Trimble had also cited the Manhattan Project as the inspiration for his proposed anti-gravity programme. Brown, too, had offered a highly unusual development model for Winterhaven, his Mach 3 anti-gravity programme, one that only peripherally involved the US aerospace industry.

I tried to keep my mind on the facts. The essential point of Twining's memo was that the development of an aircraft exhibiting the characteristics of the saucers that people were seeing in 1947 was 'within the present US knowledge', but that the craft themselves were nothing to do with the USAAF. Then whose were they?

Twining himself offered two possibilities. First, that they were of 'domestic origin – the product of some high security project not known to AC/AS-2 (air force technical intelligence) or this Command.' And second, 'that some foreign nation has a form of propulsion, possibly nuclear, which is outside of our domestic knowledge.'

The first part I actually found a little scary. Twining was entertaining the possibility that these craft had been developed and fielded by another branch of the US armed forces – the Navy, perhaps – without the knowledge of the USAAF.

The second explanation was, at least, a little more conventional. The discs, Twining offered, were the result of another country's secret development effort.

I had been sucked into a story that seemed to have had its origins in 1956. Reluctantly, I now had to concede that if I was to do this thing any justice at all, I needed to cast the net back even further.

Chapter 4

History doesn't relate the mood of the crewmembers of the Northrop P-61 Black Widow patrolling the sky above the Rhineland that night, but the available clues suggest that they were none too happy. A night-fighter crew was about as close-knit a unit as you could find amongst the front-line squadrons ranged against Germany in late 1944. Success – their very survival when it came down to it – depended on a high level of trust, intense training, the reliability of a piece of hardware then still in its infancy – radar – and undiluted concentration.

The last thing US Army Air Forces Lieutenant Ed Schlueter would have needed that night was a passenger. Worse, Ringwald wasn't even aircrew, but an intelligence officer.

Lt Fred Ringwald was sitting behind and above Schlueter, the P-61's pilot, in the position normally occupied by the gunner. Schlueter's unit, the 415th Night Fighter Squadron of the US 9th Army Air Force, had recently upgraded from their British-made Bristol Beaufighters to the Black Widow. They had also just transferred from the Italian theatre of operations to England and thence across the Channel, deploying eastwards in short hops across north-west France as the Allies pushed the Nazis towards the Rhine and back into Germany itself.

Schlueter flew his aircraft south along the Rhine, looking for 'trade'. The Black Widow was a heavy fighter; bigger than the Beaufighter and considerably more menacing in appearance. Whilst its primary targets were the German night fighters sent up to intercept British bomber streams heading to and from Germany, there was always the chance, provided Schlueter was sharp-eyed enough, of hitting a Nazi train or a vehicle convoy, especially as the Germans were moving men and materiel under cover of darkness due to the Allies' overwhelming daytime air superiority. This had been pretty much indisputable since the D-Day landings five months earlier.

But night strafing operations brought their own hazards. Over the uncertain territory of the Rhineland, sandwiched between the bluffs of the wide and winding river and the rugged uplands of the Black Forest, there was a better than even chance you could follow your own cannon-fire into the ground.

There was no official record as to why the intel officer Ringwald was along for the ride, but knowing something of the way in which spooks had a habit of ring-fencing intelligence from the people who needed it most, I had a feeling that Schlueter and his radar operator would have been left just as much in the dark.

In the black skies above the Rhineland, after a long period of routine activity, it was Ringwald who broke the silence.

'What the hell are those lights over there?' he asked over the RT.

'Probably stars,' Schlueter said, concentrating on his instruments.

'I don't think so,' Ringwald replied. 'They're coming straight for us.'

Now, Schlueter looked up and out of the cockpit. In the pitch-black of his surroundings, the formation of aircraft off his starboard wing stood out like a constellation of tiny brilliant suns. Instinctively, he twisted the control column, bringing the Black Widow's four cannon and four .50 cal machine guns into line with them. At the same time, he called ground radar.

The ground-station was supposed to be Schlueter's eyes and ears at all times, but if each pulsating point of light represented the exhaust plumes of a German night fighter, there were anything up to ten aircraft closing on him and he hadn't heard a whisper out of them. Somebody had screwed up. Later, he would be angry. Now, he was simply frightened and confused. He urgently requested information.

'Negative,' the reply came back. 'There are no bogies in your sector. You're on your own.'

Schlueter's radar operator, Lieutenant Don Meiers, who was crouched over the scope of the SCR540 airborne intercept (AI) radar in a well behind Ringwald, told him the same thing. The sky ahead of them was empty of any air activity.

But the lights were still there and they kept coming. Instead of running, Schlueter boosted the throttles and pointed the nose of the Black Widow at the lead aircraft of the formation.

As the twin air-cooled radials powered up either side of him, the glow from his opponents' exhausts dimmed; and then they winked out. Puzzled and alarmed at having lost the contacts, and with no help from Meiers on the Black Widow's own radar, Schlueter held the aircraft steady and the crew braced themselves for the engagement.

Schlueter eased the night fighter into the blacked-out bowl of sky where he had last seen the contacts. He craned his neck for a glimpse of something – anything – that signalled the presence of another aircraft, jinking the plane left and right to check his blind-spots.

Still nothing.

It was only when he started to execute a turn back to base that Ringwald told him the lights were back again.

Schlueter followed the line indicated by Ringwald and saw them a long way off. Impossibly far, in fact, but still within radar range. He called out to Meiers, but the radar operator was now having technical problems with his AI set.

Schlueter again prepared to engage the enemy, but the lights had already begun to glide away to the north-east, eventually retreating deep into the German lines and disappearing altogether.

Nobody said anything until shortly before the Black Widow landed. Although Schlueter and Meiers were agreed that the Germans must have been experimenting with some new kind of secret weapon, neither wanted to hazard a guess as to what this weapon might have been. There was nothing they knew of in their own inventory that approached the weird, darting performance characteristics of the aircraft they'd just seen.

Fearful they would become the target of unwelcome squadron humour – along predictable lines they were 'losing it' – they decided not to report the incident. And Ringwald, the spook, went along with it.

Reports of this incident exist in a number of UFO books – books I'd not encountered before because to someone steeped in the dry reportage of nuts-and-bolts technical journalism, they'd never entered my orbit.

The incident showed that almost three years before Twining wrote his memo to Gen Schulgen unconventional aerial objects had appeared in German skies prior to their manifestation across the USA in 1947.

On odd days off work and at weekends, I'd begun trawling public archives for corroborating evidence of this sighting. What I found were details on the 415th Night Fighter Squadron and the aircraft Schlueter had flown at the time of the encounter; details that allowed me to fill in the gaps of the published account and to visualise the sense of bewilderment and fear that Schlueter and his crew would have experienced that night. But along the way, I discovered that Schlueter's sighting was far from unique. All that winter of 1944–45, Allied aircrew reported small, ball-shaped aircraft glowing orange, red and white over the territory of the Third Reich. While some attributed the lights to natural phenomena such as ball-lightning or St Elmo's fire, others could not dismiss the sightings so easily. The devices appeared to be able to home in on Allied aircraft as if guided to them remotely or by some in-built control system.

Bit by bit, the reports entered the realm of officialdom. In archives and on the Internet, I found dozens of them.

'At 0600, at 10,000 feet, two very bright lights climbed toward us from the ground,' another pilot from the 415th told intelligence officers after an encounter on 22 December, near Haguenau, close to where Schlueter, Meiers and Ringwald had been. 'They levelled off and stayed on the tail of our plane. They were huge bright orange lights. They stayed there for two minutes. On my tail all the time. They were under perfect control. Then they turned away from us, and the fire seemed to go out.'

Although their appearance was sporadic, aircrews increasingly reported the devices via the appropriate channels. They nicknamed them 'foo-fighters', a bastardisation of the French word 'feu' for 'fire' that had worked its way into a cartoon strip called *Smokey Stover, the Foolish Foo Fighter* which had first appeared in a Chicago newspaper several years earlier. Meiers, who was a resident of Chicago, appears to have been the first person to have coined the term.

The consensus was that foo-fighters were Nazi secret weapons of some kind, but mighty strange ones, since they did not open fire on Allied aircraft, nor did they explode on proximity to them. They simply appeared, tagged along for a while and then vanished.

Seemingly, the spooks couldn't provide any plausible explanation for what they were either, as the following account, by Major William Leet, a B-17 pilot attached to the US 15th Air Force, indicated after a night-time encounter with a foo-fighter – 'a small amber disc' – that followed his bomber all the way from Klagenfurt, Austria, to the Adriatic Sea in December 1944. 'The intelligence officer that debriefed us stated it was a new German fighter but could not explain why it did not fire at us or, if it was reporting our heading, altitude and airspeed, why we did not receive anti-aircraft fire,' he reported.

Most encounters were at night, but there were daylight sightings, too.

A B-17 pilot, Charles Odom, flying on a daylight raid into Germany, described them as being 'clear, about the size of basketballs'. They would approach to within 300 feet, 'then would seem to become magnetized to our formation and fly alongside. After a while, they would peel off like a plane and leave.'

A P-47 fighter pilot also reported seeing a 'gold-coloured ball with a metallic finish' west of Neustadt in broad daylight, whilst another saw a 'three to five feet diameter phosphorescent golden sphere' in the same area.

In 1992, researchers digging into the foo-fighter mystery uncovered a wealth of buried reports within the US National Archives at College

Park, Maryland. What was intriguing was that almost all of them had been filed by pilots and crewmembers of the 415th Night Fighter Squadron. Unlike the incident involving lieutenants Schlueter, Meiers and Ringwald (whose experience was relayed after the war by a former war correspondent), the 15 'mission reports' of mysterious intercepts, many of which occurred in a triangular sector of the Rhineland bordered by an imaginary line linking Frankfurt-am-Main to the north, Metz to the west and Strasbourg to the south, could be examined in their original format; logged in the dispassionate shorthand of the intelligence officers who originally noted them down.

'December 22/23, 1944 – Mission 1, 1705–1850. Put on bogie by Blunder at 1750 hours, had AI contact 4 miles range at Q-7372. Overshot and could not pick up contact again. AI went out and weather started closing in so returned to base. Observed two lights, one of which seemed to be going on and off at Q-2422.'

And another:

'February 13/14 1945 – Mission 2, 1800–2000. About 1910, between Rastatt and Bishwiller, encountered lights at 3000 feet, two sets of them, turned into them, one set went out and the other went straight up 2–3000 feet, then went out. Turned back to base and looked back and saw lights in their original position again.'

The reports made it clear that the sightings covered a period between September 1944 and April 1945.

September. Two, maybe three months before Schlueter's and Meiers' own encounter.

Maybe, I thought, that tied up one little loose end.

When the spook, Ringwald, had been riding with them that night he must have been looking for something he already knew to be there.

So what did we have here?

Ostensibly, these objects – they could hardly be described as 'aircraft' – exhibited characteristics similar to T.T. Brown's flying discs. Their historical interest lay in the fact that their 'existence', if you could call it that, had been logged by observers with impeccable credentials more than three years before the first rash of flying saucer sightings in the United States. True, it was wartime and things got misidentified. But with the foo-fighters there was none of the hysteria that accompanied the US-based sightings of 1947, which tended to make the testimony, if anything, more objective and believable. These people, many of them hardened by years of combat experience, felt they were encountering something new and dangerous in German skies. And the year 1944 seemed to be key.

Their experiences echoed what USAAF Gen Twining had told Gen Schulgen in his secret memo of September 1947. That 'aircraft' making no apparent sound, with metallic or light-reflecting surfaces, exhibiting extreme rates of climb and manoeuvrability, 'were within the present US knowledge'.

What did that mean exactly?

I didn't know, but somehow I felt the wording was key.

Over the next few months, life at *JDW* went on pretty much as it had done since I'd joined the magazine in the mid-80s; a routine made up of press conferences, air shows, defence exhibitions and weekly deadlines. But as I maintained a close watch on developments that were shaping the defence and security climate of the post-Cold War world, I found my mind on other things. Things I kept to myself.

Had the Germans developed a totally new form of propulsion, blended it with a radically different kind of air vehicle, and deployed it in the form of some new and secret weapon system in the latter stages of the war? I had to juggle this possibility with the fact that T.T. Brown had postulated the basis of an anti-gravity propulsion system as far back as the 1920s.

I decided to discover whether any of this wartime sightings data was reflected in work the Germans had been doing in their research facilities and factories.

In the late 1950s, a book written by a German who had served as the commanding officer of a German Army technical unit in the Second World War, Major Rudolf Lusar, went on to become an unlikely bestseller in Britain and the United States.

The book was called *German Secret Weapons of World War Two*.

Seeking clues to the foo-fighter mystery, I discovered a copy in the reading room of the Imperial War Museum. In it, Lusar described in meticulous detail, in language that often made the depths of his bitterness clear, the technical achievements of 'a small, industrious and honest nation which lost the war'.

Secret Weapons made sombre reading. Although German technical achievements were visible in developments such as the V-1 flying bomb, a direct forerunner of the modern-day cruise missile, and the V-2 ballistic missile, it was the vast extent of Germany's underpinning technology base, as revealed by Lusar, which showed just how far ahead of the Allies the Nazis had been in certain key areas.

Jet engines, rocket engines, infrared and thermal-imaging systems, proximity fuses, missile guidance seekers . . . technologies that are integral to most modern aircraft and airborne weapon systems were all listed and described. In the late 1950s, when Lusar's book first appeared, these technologies were still in their infancy in Britain and America.

Yet the Germans had been working on them a decade and a half earlier.

But there was another side to the book, one which was so sensational that immediately on its appearance it had set alarm bells ringing in Washington.

This side of the book related to so-called German 'wonder weapons' beyond the V-1 and V-2.

One of these was the *Fleissiges Lieschen*, or 'Busy Lizzie', a 150 m-long tube assembly of non-alloyed cast steel with chambers leading off it that lent it the appearance of a giant millipede. It was designed to fire 150 mm projectiles over distances of 170 km, more than enough to hit most British cities from sites deep within France. It never went into service, but 45 years after the war the components of an almost identical weapon, the 'supergun', were intercepted by British customs officials on their way to Iraq.

Other esoteric developments detailed by Lusar showed that the Germans had been working on technologies for bringing down Allied aircraft with sound waves, air vortices, intensely focused beams of light and jets of compressed air.

In 1958, the US Air Force commissioned a 'special studies group' within Air Force Intelligence headed by an Austrian-born technical consultant called Dr Stefan Possony to carry out a detailed appraisal of Lusar's book. The research effort was branded 'secret' and has only recently come to light. I found it during a long, hard night's trawl in the basement when it popped out of cyberspace after an unusual combination of word commands fed through a high-power search engine.

A section in Lusar's book was devoted to 'flying saucers' which he asserted, in no uncertain terms, were the product of German wartime inventors. 'Experts and collaborators in this work confirm that the first projects, called "flying discs", were undertaken in 1941,' Lusar wrote. He even went on to name the key individuals involved. These were 'the German experts Schriever, Habermohl and Miethe, and the Italian Bellonzo'.

Lusar described the craft in some detail. There were two principal centres of disc development: one, headed by Miethe in the vicinity of the Lower Silesian city of Breslau in modern-day Poland; the other centred on Prague in Czechoslovakia, then also an integral part of the Reich. The Miethe disc was described as a discus-shaped 'plate' of 42 m diameter, fitted with 'adjustable' jet engines. Shortly before the plant where the craft was being built was overrun by the Russians, the retreating Germans blew it up, destroying the disc inside. However, many of the 'experts' who had worked on the project were captured and taken back to Siberia, where work on them 'is being successfully continued', Lusar reported.

In the paranoid US security climate of the late 1950s, it was this aspect of the account that had led to the commissioning of Possony's special intelligence report.

According to Lusar, the other disc, developed by Schriever and Habermohl, did achieve flightworthy status before the end of the war – on February 14, at a facility just outside Prague. 'Within three minutes they climbed to an altitude of 12,400 m and reached a speed of 2000 km/h in horizontal flight,' the author related.

At a time when the most advanced interceptors of the day (the late 1950s) struggled to achieve such velocities, this was an outrageous claim, but then it didn't have to be taken wholly at face value to contain an element of truth.

Having covered the Stealth Fighter story in the mid-1980s, when the aircraft was still deeply secret and witnesses were reporting things they couldn't square with any known aircraft development of the day, I trusted the old maxim about smoke and fire.

And in this assumption, I had an ally in Dr Possony. Why else had he been tasked by the US Air Force with writing a classified critique of Lusar's claims? Why else, according to one source I spoke to about him, a researcher by the name of Joel Carpenter who had devoted considerable time to studying Lusar and Possony, had the good doctor shredded his conclusions on the subject?

So, alongside wartime pilots' reports of flying objects that defied their understanding of aircraft performance were accounts of Germans having worked on projects that had no engineering counterpart, even within present-day aerospace knowledge. In Lusar's book were names, dates and places – crude data, admittedly, but usable nonetheless as jumping-off points for a more rigorous search. And though Lusar's claims didn't point directly to evidence of anti-gravity propulsion technology, in his discussion of another taboo technology area in scientific circles, the flying saucer, he had outlined a pathway to an altogether different kind of aerospace platform.

Just as T.T. Brown had done in the 1920s – and Trimble had in 1956.

The task seemed so simple. Make the link between Germany and the flying saucer and here was an opportunity to solve not only the anti-gravity propulsion riddle, but, in the process, perhaps, one of the most baffling mysteries of the 20th century: the origins of the UFO. If machines like those in General Twining's memo were 'within the present US knowledge', no wonder Trimble and his colleagues had so quickly fallen silent on the subject of a new and exotic propulsion source.

The flying disc must have exhibited performance so in advance of its time that it had been super-classified, then hidden in plain sight – behind the UFO myth – for the best part of 60 years.

Perhaps, too, it explained why Trimble had sounded like he'd had the fear of God put in him when he had been approached by Lockheed's PR machine on my behalf. The spooks would have made it abundantly clear there would be no statute of limitations on this particular secret. That you never talked about it. Ever.

Across multiple satellite relays, I heard the anguished shrieks of his three-month-old, the background clatter of his older children as they ran wild in the tiny apartment and finally relief in Lawrence Cross' voice as he realised it was me; a moment's respite in a whirl of copy deadlines, late night feeds and trips to theme-parks.

The baby's crying faded into the static as Cross took the handset into another room.

I thanked him for the tip about LaViolette and Valone, as a result of which I had pieced together the essential elements of T.T. Brown's life and work. Without elaborating on how I'd arrived there, I asked him what he knew about the Germans' purported development of flying saucers in World War Two.

'Oh Jesus,' Cross said, and I formed a mental picture of him rubbing his eyes and reaching for a cigarette, 'where do I begin?'

I prompted him. One minute the Nazis were tinkering with this technology, the next people were reporting sightings of these objects, first in Gemany, then in America . . . ?

'It's not as simple as that,' Cross said.

'Come on, Lawrence. A book called *German Secret Weapons of World War 2*, by Rudolf Lusar details names, dates and places. Has anyone ever dug into this stuff? Have you? The story must be fit to bust wide open.'

'It's been around for decades,' he said, 'long enough to have been given a name.'

'What do you mean?'

'In the trade, we call it "the Legend". It looks so straightforward, doesn't it? A story with a solid trail. But it's not like that at all. When you get into this stuff the trail goes everywhere and nowhere. The people who really did exist are long dead and others probably never existed at all. I know. I've been there and I've looked for them. So have dozens of other investigators. The detail is fantastic, but it's all uncorroborated. By that I mean there isn't a shred of evidence in any archive – no official word, no plans, nothing – that any of these projects ever existed.'

'Have you got anything on it?'

He laughed and it caught in his throat. Cross smoked for the entire journalistic profession. He hacked the cough into the room, away from the receiver.

'Sure. How much time have you got? I'll email it to you. There's another book you should get a hold of. It's by an Italian. Guy by the name of Vesco – Renato Vesco. The book's called *Intercept But Don't Shoot*. There's an English translation. I think it was first published in the late 1960s. Vesco's well up there, right in the heart of the Legend. Maybe you should try and track him down – that is, if he really existed.'

I said nothing. Cross continued.

'Take nothing at face value and you're off to a better start than me. *Lusty* was real enough, but that's all I know. Everything else is up for grabs.'

I thought I must have heard him wrong. '*Lusty*? Did you say, *Lusty*?'

'It stands for *Luftwaffe Secret Technology*. The US Army Air Forces' official file on the state of the German Air Force's secret weapons work at the end of the war. The Air Force Historical Research Agency has a copy in its archives at Maxwell Air Force Base. It's raw data. That is to say, it's never been processed by anyone, so it's uncorrupted and verifiable, but I have no idea what's in it. In all these years, doing what we do, I've never had an excuse to go there. Maybe you do.'

'Where is Maxwell Air Force Base?'

'It's in Alabama. Where the skies are so blue. A distant memory for you, I guess.'

I thanked him, hung up, and knew that I wouldn't be tapping Cross for any more advice. He'd made it abundantly clear, more in tone than anything else, that for him this was the end of the road.

I thought about the months ahead. I had no legitimate business in the States except for a trip to Washington to cover an Air Force symposium – and that wasn't until later in the year. Stretching the trip to include a visit down to Alabama was simply out of the question.

I called the Air Force Historical Research Agency at Maxwell Air Force Base to see if I could get a copy of *Lusty* sent to me. In theory I could, but the files were held on a number of microfilm reels and, I was told, it could take months, perhaps years, for the release forms to plough through the system and for the report to reach me. Not only that, but the

Lusty files were old and corrupted, making repros difficult. The best, probably only way to view them, the administrator there told me, was in situ.

I put the phone down, kneaded my eyes and thought about smoking a cigarette again; something I hadn't done in years.

I had just hit another pluperfect dead-end.

Chapter 5

Flugkapitän Rudolf Schriever, one of the four saucer engineers cited in Lusar's book, began talking to the West German media in 1950 about a truly fantastic flying machine he had worked on for the Nazis – one that would have changed the course of the war had it gone into full-scale production. It stemmed from work that he had allegedly undertaken for the Heinkel Aircraft Company at Marienehe, near Rostock on the Baltic coast.

Schriever was one of the four 'scientists' mentioned in Lusar's book as having worked on the Nazi flying discs. He was also, I found out through the massively detailed file that Cross sent me via email, central to the Legend.

Though he started out as a pilot in the Luftwaffe, Schriever appeared to have developed some highly advanced ideas about aircraft that could take off and land vertically and it was in this capacity, after he had been drafted to Heinkel's design section, that he soon came to the attention of company chairman Professor Ernst Heinkel, who in early 1940 encouraged him to construct a small flying prototype.

So far so good. Though best known for its lumbering He 111, the Luftwaffe's mainstay bomber during the Blitz against London in 1940 and for much of the rest of the war, Heinkel was one of the most pioneering and innovative aircraft companies within Germany at the time. Whatever the Legend said, this much was fact.

In 1936, Ernst Heinkel began funding experiments that three years later would lead to Germany's successful construction of the world's first jet-powered aircraft, despite the fact that it was the British designer, Frank Whittle, who had first invented and patented the concept. The tiny, one-off Heinkel He 178 first flew on 27 August 1939, five days before the German Army marched into Poland. Eighteen months later, Heinkel would again eclipse all other aircraft companies by flying the world's first jet-powered fighter, the He 280. If anyone was to develop something as radical as Schriever's idea, therefore, Heinkel was the company to do it.

The Legend then takes over.

In the spring of 1941, Schriever's blueprints were being used to construct a 'proof-of-concept' model in a 'garage' away from prying eyes. Officially known as the V1 (V for 'versuchs' or 'experimental version 1'), informally it was referred to as the 'Flying Top'. It was probably no more than two or three feet in diameter and powered by an electric motor or a small two-stroke engine. It is not known where this garage was, but within the sprawling complex of buildings at Marienehe, set in three square kilometres of the Mecklenburg State Park, there was ample room for Schriever's esoteric little engineering project to be hidden from view. Furthermore, it fitted into the Heinkel way of doing things. During the development of the world's first turbojet, Ernst Heikel had installed Dr Hans-Joachim Pabst von Ohain, a gifted graduate of the University of Göttingen, in the same kind of environment – a converted garage at the university – before transferring the fruits of von Ohain's labours, the revolutionary HeS 3A turbojet, to a secure facility within the Marienehe site.

By June 1942, Schriever's Flying Top had been test flown and the results deemed sufficiently interesting to secure top-secret funding from the RLM, the Reichsluftfahrtministerium or State Air Ministry.

With RLM funding, the intention was to construct a full-scale piloted version capable of controlled vertical take-off and landing. Construction of this full-size version, the V2, began at Marienehe in early 1943. The V2, which was known as the 'Flugkreisel' or 'Flightwheel', had a diameter of approximately 25 feet, its power generated by one or perhaps two Heinkel-Hirth jet engines, depending on which version of the Legend you want to believe. The V2 supposedly flew with Schriever at the controls, but as a piece of technology it was deemed to be heavily over-engineered and was quickly scrapped. As a proof-of-concept vehicle, though, it seems to have served its purpose, because shortly afterwards, Schriever and his team relocated to Czechoslovakia where they set about constructing a larger and altogether more sophisticated prototype known as the V3.

With the Allied aerial bombing campaign now at its height, their activities were dispersed around the Prague area to minimise the exposure to the relentless air attacks, by now penetrating deep into the Reich. But the bulk of the team's work was centred on a restricted area of a satellite facility outside Prague belonging to the Munich-based Bayerische Motorenwerke engine company, better known today as BMW. Despite the existence of Heinkel's own jet engines, the real cutting edge of German gas-turbine research was centred on BMW and in particular on its Bramo division, located at Spandau in Berlin. Bramo, the Brandenburgische Motorenwerke, had been bought by BMW from Siemens in 1939. By the middle of the war, BMW–Bramo had 5,000 staff working full-time on gas-turbine research alone – a discipline then barely a decade old – and was ultimately responsible for the BMW 003 jet engine, the best turbojet of the day, which powered the Me 262, the world's first operational jet fighter, and the Ar 234, the Luftwaffe's advanced jet-propelled reconnaissance-bomber, another aircraft far ahead of its time.

It was from Spandau, supposedly, that Klaus Habermohl, the second disc engineer mentioned by Lusar, was recruited to the Schriever team. Habermohl's job in Prague was to integrate the disc with a new and radical form of powerplant called the radial-flow gas-turbine, or RFGT. Unlike Brown's electrogravitic motor, the RFGT was, at least, recognisable technology by modern standards, if extraordinary. It was essentially a jet engine. However, unlike a regular jet powerplant, with its compressors, combustion chambers and turbines mounted one behind the other in what was basically a big tube, the RFGT formed part of the airframe itself, with the whirling turbo-machinery rotating around the aircraft's centrally mounted cockpit. As such, 'aircraft' did not adequately describe what the machine actually looked like. There was only one configuration to which an RFGT could possibly be adapted: that of a flying disc or saucer.

By the autumn of 1944, the V3 is said to have been completed. With German airfields under constant attack from Allied daylight bombing, the prospect of a fighter or bomber that could take off and land vertically from any dispersed site would have been exactly in line with Luftwaffe requirements. However, due to 'an administrative change', the V3 programme was abandoned in favour of a further prototype, the V7, propulsion coming from another experimental RFGT from BMW–Bramo. The V7 supposedly had a diameter of 60–70 feet and a crew of two or three.

Applying a crude rule of thumb, based on a rough estimate of the disc's weight, Habermohl's ingenious RFGT would have had to have generated around 10–15,000 lbs of thrust to have made the design of the disc in any way viable. This was the machine that supposedly test-flew on February 14 with Schriever and Habermohl at the controls, achieving 2,000 km/h in level flight.

The fastest aircraft of the day, the little rocket-powered Me 163, struggled to attain half this speed.

Lusar's third and last German saucer scientist, Dr Richard Miethe, was supposedly working on another disc project at a subterranean facility near Breslau under the auspices of an altogether separate contract. Towards the end of the war, the legend stated that Miethe was drafted from his activities in Breslau to assist with the Schriever/Habermohl disc in Prague, an indication, perhaps, that the Schriever/Habermohl disc was the better bet in a procurement environment that was by now desperately short of money, skilled labour and raw materials. This streamlining coincided with the 'administrative change' that had led to the abandonment of Schriever's V3 design for the altogether more viable V7.

The V7, then, seems to have been the result of a three-way endeavour, although there were reports that a Miethe disc – based, perhaps, on the project he abandoned at Breslau – was captured by the Russians, along with a number of engineers and scientists, when they drove the Germans out of Poland.

As for the V7, some say it, too, was acquired by the Russians when they took Prague; others that it was blown up by the Waffen-SS on 9 May 1945, the day hostilities in Europe ended. The Legend had it both ways.

The detail in the Schriever/Habermohl/Miethe legend was rich and impressive. Here were names, dates and places – minutiae even, that seemed to corroborate everything Lusar had written down. The trouble was, the data was based heavily on the say-so of Schriever, who was long since dead; the rest had magically appeared out of thin air, just as Cross had said. No one knew where the detail emerged from. Over the years it had been passed down from one researcher to the next, with no apparent attribution. When I approached BMW's archivists, for example, they denied that there had been any BMW factory near Prague 'engaged in advanced aircraft projects including design or advanced research during World War 2.'

I began to see what Cross meant. Cross and I were used to sourcing every piece of information we ever came by. This stuff was unverifiable. What I needed was hard, solid proof; and within the Schriever portion of the Legend there was nothing to hang a hat on, beyond the fact that the man himself had existed.

Unfortunately, the same could not be said of the other characters in the story. All attempts by researchers to trace Miethe and Habermohl had foundered, although there had been occasional 'sightings' in the lore that had grown up around them.

Miethe is supposed to have escaped Czechoslovakia in early May and to have headed west, eventually making contact with US technical intelligence teams operating inside Germany. Herded into a 'pen' along

with Wernher von Braun and his fellow rocket scientists, Miethe was said
to have been taken to the United States, ending up at Wright-Field, the
USAAF's premier research and development centre near Dayton, Ohio.
If any of this actually happened, there is no trace of it.

Habermohl is said to have been captured by the Russians at the Letov
factory, a German-administered aircraft plant outside Prague, and, after
a period of detention, sent to work at a top-secret Soviet aircraft design
bureau east of Moscow.

Again, no one could say for sure whether he had even existed at all.

My research showed the Italian, Bellonzo, was real enough, except
Lusar had misspelled his name. In 1950, Professor Giuseppe Belluzzo, a
former industry minister in Mussolini's cabinet, started talking about
disc-shaped 'flying bombs' that he claimed to have worked on during the
war and passed on to the Germans, who had subsequently developed
them into working prototypes.

Belluzzo was also convinced that these weapons were the basis of the
flying-saucer sightings that had gripped much of America for the best
part of three years and that they were now under further development
inside the Soviet Union. Beyond the fact that Belluzzo, like Schriever,
was real, his claims also remain unverifiable. Interestingly, though, he
started talking to the media just a few days before Schriever, leading
some researchers to reason that Schriever, a man who like so many other
Germans in 1950 was struggling to make ends meet, had invented his
entire story.

In fact, the Legend lacked a single item of corroborating data. There
was nothing in any archive or museum, no photograph, no indisputable
piece of testimony, to say that any of it was true.

There were splits and schisms in the Legend, just as there are orthodox
and unorthodox branches within major religions. The other strand to the
myth that I had to pay attention to was Vesco's. I had managed to obtain
a copy of his book, *Intercept But Don't Shoot* (published 1971), and
scrutinise it. What was beguiling about Vesco's account was the certainty
with which he presented his case. Vesco, who was 22 years old when the
war ended and said to have been well connected with technical experts
within the Italian Air Force at the time he wrote *Intercept*, claimed that
he pieced together his account of the top-secret development effort
behind the foo-fighter programme from sources inside the Italian
military and from Allied intelligence reports published after the war.

What also made Vesco's account worth more than a cursory glance was
the fact that it 'detailed' a completely different development effort from

the Schriever/Lusar account. Vesco claimed that there were two kinds of foo-fighters. One was unpiloted and remotely controlled; in effect, a flying bomb designed to knock down enemy aircraft. The other was manned, but flew in anger only once before hostilities ended.

The unpiloted version was called the Feuerball (Fireball) and it blended a number of highly advanced technologies.

'In the autumn of 1944,' Vesco wrote, 'in Oberammergau in Bavaria, the OBF – an experimental centre run by the Luftwaffe – had completed a series of researches into electrical apparatus capable of interfering with the operation of an engine up to a maximum distance of about a hundred feet by producing intense electromagnetic fields.'

In parallel, a separate effort was under way by the Germans to produce a 'proximity radio interference' device capable of jamming or spoofing Allied radio and radar systems. Put these two technologies into a small, circular, armoured airframe 'powered by a special turbojet engine, also flat and circular and more or less resembling the shell of a tortoise' and 'a highly original flying machine was born.'

Radio-controlled at the moment of take-off, the machine was steered towards Allied bomber streams by a ground-operator, whereupon it automatically latched onto their slipstreams, 'attracted by their exhaust flames, and approached close enough without collision to wreck their radar gear.'

And then came the detail. The Fireball was first constructed at an aircraft plant at Wiener Neustadt, south of Vienna, with the help of the Flugfunk Forschungsanstalt of Oberpfaffenhoffen (FFO), an aircraft electronics firm near Munich. Hermann Goering, Hitler's deputy, inspected progress on the weapon 'a number of times', hoping that the principle of the Fireball could also be used to produce 'an offensive weapon capable of revolutionizing the whole field of aerial warfare.

'The fiery halo around the perimeter – caused by a very rich fuel mixture – and the chemical additives that interrupted the flow of electricity [in Allied aircraft] by overionizing the atmosphere in the vicinity of the plane, generally around the wing-tips or tail surfaces, subjected the H2S radar on the plane to the action of powerful electrostatic fields and electromagnetic impulses (the latter generated by large klystron radio tubes protected with anti-shock and anti-heat armour). Since a metal arc carrying an oscillating current of the proper frequency – equal, that is, to the frequency used by the radar station – can cancel the blips (return signals from the target), the Feuerball was almost undetectable by the most powerful American radar of the time, despite its night-time visibility.'

Once again, I pictured the desperation felt by Schlueter as his radar operator, Meiers, failed to register the orbs of light ahead of the Black Widow on the SCR540.

Was this what they had encountered that night, the Fireball?

To believe that it might have been, I had to accept that what we had here was a weapon system that was decades ahead of its time. Years before the deployment of the surface-to-air guided missile – sufficiently perfected to shoot down the CIA U-2 pilot Gary Powers over Russia only in 1960 – and decades before radar-evading stealth technology would become a household word, the Vesco portion of the legend held that the Germans had developed a single weapon system that blended these and other exotic technologies. Not only that, but the Fireball was able to home in on its intended prey, the Americans' B-17 and B-24 bombers, via an automatic guidance system – one that tracked their exhaust plumes – a sensor so advanced that it was unheard of, even with the benefit of modern-day knowledge. Once close to the bombers, it could then either disrupt their electronics or stop their engines in flight with a gas that killed their ignition systems.

At a stroke, Vesco had assembled all the evidence he needed to explain the foo-fighter conundrum. He also threw in some eye-witness testimony. 'One person who saw the first short test flights of the device, without its electrical gear,' he wrote, 'says that "during the day it looked like a shining disc spinning on its axis and during the night it looked like a burning globe."'

The trouble was, it sounded like science-fiction.

When the Russians advanced into Austria, Vesco maintained, the Fireball production line was moved from Wiener Neustadt into underground facilities run by the Zeppelin Werke in the Schwarzwald – the same area of Rhineland and Black Forest where Schlueter, Ringwald and Meiers, and a host of other aircrew from the 415th Night Fighter Squadron, had first encountered the inexplicable.

Vesco went on to explain that the Feuerball had an 'older brother', the Kugelblitz or 'Ball Lightning Fighter', that was built and flight-tested – once – in the vicinity of Kahla, site of a huge underground weapons development complex in Thuringia, a region of mountainous uplands in the heart of Germany known as the Harz. If this 'fact' also was a hoax, it was a clever one, since the Harz certainly had housed a number of underground Nazi weapons factories, amongst them the facility at Nordhausen, where Wernher von Braun's V-2 rockets had been produced.

Vesco was notoriously reclusive. Some researchers believed he'd never existed at all; lending credence to the supposition that *Intercept* really was just an elaborate hoax. But Vesco was real, all right. I managed to trace a group of Italian researchers who had been in contact with him up until his death in November 1999. According to them, Vesco didn't care whether people believed him or not.

'He was very correct,' one of them told me over the phone, 'so it's very hard to believe that he made it all up.'

But where was the proof?

Vesco claimed that the evidence for Feuerball and Kugelblitz was to be found within obscure tracts of the British Intelligence Objectives Sub-committee (BIOS) reporting system and its successor, the US-UK administered Combined Intelligence Objectives Sub-committee (CIOS), published in the immediate aftermath of the war (a number of reports from which, however, are still withheld). BIOS and CIOS were the systems employed by the British and the Americans for assessing German high technology. But researchers had been through all the available CIOS and BIOS files with a fine-tooth comb and had found nothing that pointed to anything that described the Fireball or its 'older brother' the Ball Lightning Fighter.

Vesco had stuck to his guns, never wavering from the conclusions he had espoused in his book: that the Germans had developed a truly revolutionary new form of air vehicle; that it was the British who had happened upon the technology at the end of the war and that they, together with Canadian scientists, had refined it in the frozen wastelands of British Columbia and Alberta.

Vehicles resulting from this endeavour, Vesco maintained, were responsible for the rash of US flying saucer sightings in 1947.

A familiar pattern was emerging and it wasn't helping my case. The more I looked, the more 'evidence' I found that the Germans had been tinkering with technology that explained the foo-fighters sightings of that winter of 1944–45.

From Schriever in the 1950s to Vesco's testimony in the 60s – and others before, during and since – maybe a dozen Germans had come forward to say that they had worked on flying saucer technology under the Nazis. And some, like Viktor Schauberger, were so very nearly believable.

Schauberger, whose story was contained in notes on the Legend sent to me by Lawrence Cross, was an Austrian who had supposedly invented a totally new form of propulsion based upon a principle called 'implosion'. No one I spoke to seemed entirely sure what implosion meant, but according to the stories that had grown up around this man, the implosion process was at the heart of a radical turbine that Schauberger had installed in a sub-scale flying disc sometime during the war. A test of this small flying vehicle, with its echoes of the Schriever Flying Top, had supposedly taken place and the results were said to have been highly impressive. In one account of the test, the craft had apparently risen towards the ceiling of the test facility 'trailing a glow of ionisation'. This immediately elevated the report above the many others I had come across, for it signalled that, whatever was occurring within the implosion process, it had precious little to do with jet propulsion. If true, it could only have been an anti-gravity effect.

If true.

Though Schauberger was long dead, his son Walter was still alive when I telephoned him at his home in Austria in 1991. Talking through an interpreter, I asked Walter about his father's experiments and got a disconcerting answer. I had imagined that Viktor had attended some impressive German or Austrian technical institute; that he was a professor of physics at Vienna or Salzburg or, at the very least, an eminent aerospace engineer. But no. Walter Schauberger told me that his father had developed his radical ideas about energy and propulsion by observing what he had seen in Nature, the way rivers flowed and fish swam. His only professional training had been as a forester.

This was the trouble. While all of these stories were laced with detail, which immediately gave them a veneer of credibility, they were almost always let down either by corrupted data or a complete absence of it. In a decade of investigating the aerospace industry, I had never once come across a designer who did not have a set of initials after his name. A forester was simply absurd.

Had there been more to the Schauberger story, I would not have hesitated to make the trip to Austria. But as decisions went, this was an easy one. I was using my training to make value judgments all the time on the data that was gathering in my basement. You didn't need a degree in physics, though, to appreciate that whatever they had or hadn't been, foo-fighters would have had to have come from the minds of engineers – not from a man who'd spent his time amongst Alpine forests and streams.

It was for this reason that I gently declined Walter Schauberger's offer to visit his 'biological-technical institute' in the Salzkammergut Mountains.

For any of the 'evidence' about German flying saucers to be irrefutable, it had to emerge from official documentation. But Lawrence Cross and a number of other credible researchers I'd contacted on the subject had already been through the BIOS and CIOS reports and found nothing. And that should have been the end of it. Yet, I knew that if the Germans had developed a revolutionary air weapon, based perhaps on some radical form of propulsion technology, it was inconceivable that the CIOS or BIOS intelligence teams would have documented the discovery for the world to read about.

To go the extra mile, I realised I'd have to access the original intelligence data on which the BIOS and CIOS assessment teams, most of whom were technical types sat at desks in Brussels, Paris, London and Washington, had based their analyses. Here again, I knew there would be no smoking gun. Had intelligence units in the field come across anything meaningful, these reports, too, would have been sanitised. Anything as obvious as a full-blown air vehicle or a new form of propulsion system would not have escaped the censor's attention.

But to identify evidence of a new form of aerospace technology, particularly one as far-reaching as that described in the legend or by Vesco, I wasn't searching for the obvious, because the obvious would have been picked up by the censors.

As with any new piece of technology, developments came together as a series of systems and sub-systems. There would have been prime contractors and sub-contractors, some of them working in new scientific areas – areas that would have been on the very edge of the technical knowledge of the desk-bound BIOS and CIOS assessors.

Between the chaos of the front-line and the over-stretched resources of the intelligence units tasked with assessing German hardware and documentation – which the victors had shipped out of Germany by the ton-load – there might still, I figured, be some fresh evidence; something useful that had been overlooked.

I started with the British, whose initial stab at organised technology plunder was vested in the hands of a rag-tag private army – composed, bizarrely, of sailors and Royal Marine commandos. This outfit was the brainchild of a certain Commander Ian Fleming, who 15 years later would go on to create the Bond persona in the double-oh-seven novels. Bond's character, it appears, drew heavily on the exploits of 30 Assault Unit RN, which rode rough-shod over the conventions of the day. Following the battle for Cherbourg, in which 30 AU RN had been tasked with capturing German naval headquarters, the marines liberally enjoyed the spoils of war. Their behaviour was described as that of 'merry courageous, amoral, loyal, lying toughs, hugely disinclined to take no for an answer from foe or *fräulein*.'

As a result of their 'martial exuberance', as one assessor wrote, 30 AU RN was reined in somewhat and renamed 30 *Advanced* Unit RN. It was also ordered to subordinate its activities to a new and much larger force of British tech-plunder units, known as 'T-Forces'. These would move forward with General Bernard Montgomery's 21st Army Group, principally in Monty's theatre of operations, north-west Germany, but not exclusively so. Under the terms of the CIOS charter, with its Anglo-American remit jointly to exploit German spoils of war, British T-Force teams would also be allowed to tag along with forward US Army units in pursuit of their objectives.

These objectives were to locate and secure intact technical 'targets' of interest; to preserve German high technology from 'destruction, loot, robbery and, if necessary, counter-attack', until the completion of their examination by teams of experts or until their removal. They were also to act as armed escorts in enemy territory for the 'expert investigators' drawn from CIOS offices far behind the front lines. As a quid pro quo, US technical teams could ride with forward British assault units. The eventual size of British T-Forces would grow to 5,000 personnel.

But from the reports generated by these units, now freely available in the UK Public Records Office, it was apparent that the British were desperately ill-prepared to make the most of the opportunity that lay before them.

Over the next two months, I spent every spare moment down at the Public Records Office. And when I wasn't in the Records Office, just down the road in Kew on the outskirts of London, I was gathering as much open-source reading material as I could on the Allies' systematic plunder of German high-technology at the end of the war. One book, *The Paperclip Conspiracy* by Tom Bower, read like a manual on how to dismantle an entire nation's technology base. If America, Britain and their allies had applied the lessons of *Paperclip* to the Iraqi problem at the end of the Gulf War in 1991, the threat posed by Saddam Hussein's war-machine would have been eradicated forever.

During the November 1944 to March 1945 planning stage of the T-Force operations, the problem facing British investigators was a fundamental lack of intelligence on what they were supposed to be looking for. A 'black list' of technical targets was drawn up, the majority of them weapons development and research centres put forward by CIOS.

But a general air of ignorance of the situation on the ground persisted, as a T-Force field commander later recorded. 'It appeared that the sponsoring ministries knew little or nothing about the specific whereabouts and natures of their targets; and that investigators who would eventually come out would know even less.'

If there was any sense at all that the British realised they were in a race against the other allies – the French and the Americans particularly – it came far too late. Roy Fedden, a senior and respected British aviation industrialist who flew into Germany to view the tech-plunder operation for himself, complained that his compatriots were being 'lamentably slow' to take advantage of the information on offer. With many of the UK technical experts drawn from industry, part of the problem was a considerable reluctance on their part to accept that Germany could possibly have been as far ahead of British technology in certain pioneering areas – the jet-engine, for example – as they evidently were.

There were British successes – the capture of the German navy's laboratories at Kiel, for example, where highly advanced German U-boats and torpedoes, propelled by an innovative peroxide-fuelled engine, were being developed. There were also significant British finds at Krupp in Meppen, where advanced armour and artillery technology had been produced for the Wehrmacht.

But reading between the lines of the British-controlled tech-plunder operation, its disparate character was clearly a factor in its limited success. Without adequate direction from London, there was confusion as well as a palpable lack of urgency about the operation at the front. And in some cases – the activities of T-Forces of the 21st Army Group's Canadian First Army, for example, who were roaming between Germany and Holland, looking for targets – it was portrayed as being akin to a jaunty adventure.

Among the most prominent targets on First Army's T-list, according to records of its operations at the time, was the radio-transmitter of Lord Haw-Haw, the British traitor and Nazi propagandist. 'Two further wireless stations had been discovered on the (German) island of Borkum and another platoon was dispatched there as guard. That platoon was in the happy position of having 4500 German marines to wait on them!'

It all seemed a very far cry from the exotic weaponry that Vesco had described in *Intercept But Don't Shoot*.

In scanning the British documents, however, I'd found one big problem with Vesco's thesis that the British had discovered flying disc technology and removed it to Canada. By far the greater part of Germany's advanced aerospace development had taken place in Bavaria in southern Germany, a long way from the main thrust of British operations in the north-west of the country.

And while some British T-Force units had been integrated with American forward army operations in central and southern Germany, they were more often than not outwitted by the Americans, who time and again seemed to have had the resources to take what they wanted, much to the dismay of Roy Fedden.

'The Americans,' he wrote, 'have fine-combed the country, removing considerable quantities of drawings, technical records and actual equipment direct to the States.'

This, I could see, was no mere happenstance. American tech-plunder activities were the result of the most carefully thought-out strategy in the history of US military operations, its orchestration planned at the highest levels.

In November 1944, the US Joint Chiefs of Staff had established a Technical Industrial Intelligence Committee to seek out anything in Germany that might be useful to the post-war American economy. Nor did the Joint Chiefs' target-list just comprise military objectives. One sub-committee had a staff of 380 specially trained civilians set up specifically to represent the interests of 17 US companies. Special agents from the US Field Intelligence Agencies (Technical) scoured Germany for vacuum tubes a tenth of the size of the most advanced US devices, and condensers made out of zinc-coated paper, which were 40 per cent smaller and 20 per cent cheaper than US condensers – and, instead of 'blowing' like the US vacuum tubes, were 'self-healing' – in other words, they could repair themselves. Such innovation would later prove invaluable to the post-war US electronics industry.

The teams were also on the look-out for German textile and medical advances – and found them by the ton-load. At the German chemical giant I.G. Farbenindustrie, notorious for its role in the development of the gas-chambers of the Holocaust, investigators found formulas for the production of exotic textiles, chemicals and plastics. One American dye authority was so overwhelmed by the discovery that he declared: 'It includes the production know-how and the secret formulas for over 50,000 dyes. Many of them are faster and better than ours. Many are colours we were never able to make. The American dye industry will be advanced at least ten years.' German bio-chemists had also found ways of pasteurising milk using ultra-violet light and their medical scientists had discovered a way of producing synthetic blood-plasma on a commercial scale.

Hundreds of thousands of German patents were simply removed and brought back to America.

No wonder the Brits were struggling. It was clear from these and other accounts that CIOS, the Anglo-US reporting channel and assessment office for German high-technology, was little more than a front. The real American tech-plunder operation had been organised long beforehand, under separate cover, in Washington.

A year after the war ended the US Office of Technical Services (a body set up to ensure that German technology, where required, was spun rapidly into American industry) was reportedly sifting through 'tens of thousands of tons' of documentation. This 'mother-lode' of material, according to one contemporary report, 'very likely contained practically all the scientific, industrial and military secrets of Nazi Germany.' Had exotic propulsion science been buried somewhere within this mountain of raw material, it would have been almost impossible to locate. But after the war, the mother-lode was divided between the Library of Congress, the Department of Commerce and the technical archives at Wright Field. Most of the Wright Field material had come via *Lusty*, the post-war intelligence report on Luftwaffe technology that Lawrence Cross had told me about. This was the report that was stuck, frustratingly, in Alabama – nowhere near the Air Force symposium that I was due to hit in a quickie assignment to Washington later in the month.

I called the Air Force Historical Research Agency at Maxwell Air Force Base in Alabama again to see if the administrator there could recommend any other sources of information comparable to the *Lusty* report. Since I was going to be in the vicinity of the US National Archives in Maryland, I wanted to be sure I wasn't missing out on anything.

'If you're gonna be in D.C. anyway,' she told me, chewing slowly on the syllables, 'you might want to check out the other copy of *Lusty* that's available at our sister site down by the Old Navy Yard, just across the river from Capitol Hill.'

For a moment, I thought she was joking. This was the person who'd told me, in painstaking detail, that if I wanted to obtain a hard copy of *Lusty* from the archives of Maxwell Air Force Base it would take me months or even years; that due to the poor condition of the microfilm reels, the only viable way of accessing the information was by reading the reels in situ.

When I pointed this out, her answer was nothing if not honest.

'You never mentioned nothin' 'bout no trip to D.C.,' she said.

The Office of Air Force History maintained an out-of-the-way archive in Washington D.C. located at Bolling Air Force Base, next to the Old Navy Yard.

I was supposed to be in and out of Washington inside a couple of days. I stretched the schedule and bought myself another 24 hours.

If there were trace-elements of a suppressed technology within the US documents plundered from Nazi Germany, I knew they would be buried in *Lusty*.

Chapter 6

It was still dark when I exchanged the warmth of the Holiday Inn for the damp chill of the pre-dawn, heaved my bags into the back of the cab and hunkered down for the 30-minute ride across the river. I had around eight hours to sieve *Lusty* for evidence of a technology Cross maintained couldn't be substantiated.

The commuter-traffic was jamming the bridges, the top of the Washington Monument scraping the overcast, as the taxi headed across the Frederick Douglass Memorial Bridge and into a part of the Nation's Capital that only makes the news when the body-counts from the drive-by shootings top three or four.

By the time I reported to the guard at the main gate, the wind had got up, lifting sheets of spray from a choppy confluence of grey wavecaps where the Potomac and Anacostia rivers meet off Greenleaf Point. The spray snaked in rivulets this way and that in front of us as the cab driver navigated the maze of open streets looking for the address I'd been given. It took him 15 precious minutes to find it.

The archive was tucked away in a forgotten corner of Bolling Air Force Base, home of the Defense Intelligence Agency and the USAF's Office of Special Investigations. It wasn't the sort of place you'd want to stroll or take the view.

Bolling was a large cordoned-off military district sandwiched between the black waters of the Potomac and the badland housing projects of Congress Heights and Washington Highlands. It also bordered the grounds of St Elizabeth's – a hospital for the criminally insane, since renamed – where John Hinckley served time for his assassination attempt on President Reagan.

The reading room in the tiny Office of Air Force History was low-lit and windowless. True to her word, the senior archivist had placed a cup of coffee next to the warmed-up microfilm reader opposite the furthest row of metal book stacks. There was a clock on the wall above the door to her office. It was a little after seven-thirty. The air conditioning had not

64

yet kicked in and my breath mingled as condensation with the steam rising from the cup.

The documents were contained on several reels of microfilm and, like the reels at the main repository in Alabama, they were in bad condition. My business schedule allowed for one shot at a read-through. My plane left at 6.15 that night.

To help me cut to it, the archivist had offered to open the office early and said I was welcome to turn up any time after seven. Coffee was on the house.

Time was already short when I slapped the first reel into the microfilm reader and got to work.

Lusty opened with a potted history, the opening paragraph of which began:

'At a medieval inn near Thumersbach near Berchtesgaden [Hitler's mountain-top retreat in Bavaria], early in May 1945, the German General Air Staff patiently awaited the outcome of surrender negotiations taking place in the north. They had arrived by car and plane during the past weeks, when the fall of Berlin was imminent, and had kept in contact with Admiral Doenitz at Flensburg. Through the interception of one of these messages, their location, which had previously been unknown, was discovered. Within 24 hours, Lt Col O'Brien and his small party, representing the Exploitation Division of the Directorate of Intelligence, USAFE (United States Air Forces in Europe), had arrived, located the party and conducted the first of a series of discussions with General Koller, who was then in command.'

Colonel O'Brien's men were the advance guard of 200 officers chosen from HQ Army Air Forces to oversee the USAAF tech-plunder operation and it was clear right from the start that they were in a race against time.

In the chaos of the collapsing Reich, many German scientists were dead, others had been captured by the Russians advancing from the east or the Americans, British and French in the west. Many were held in internment camps, but in a country brimming with former Reich slave-workers and displaced citizenry – almost all of whom had to be filtered by the Allies for Nazi party-members and war criminals – it was hard to know who was who or where.

But the vast majority were still at their factories and laboratories when the advance units of the USAAF plunder operation screeched up in their jeeps and half-tracks. With no orders to stop working, they had carried on at their work-benches, even though the armed forces of the high commands that had commissioned their work had either capitulated or been wiped off the map.

Many documents and blueprints were missing. German project managers would tell US investigators that they had destroyed them under orders, but the USAAF officers quickly learned to apply some psychology of their own: men and women who had devoted years of commitment to technologies they considered vastly superior to those of their enemies were incapable of such vandalism.

Most of the files existed, but had been hidden.

Germany had already been divided by the Allies into zones of occupation; and many of the most technologically interesting German facilities – the underground V-1 and V-2 production sites in the Harz mountains, for example – were in the designated Russian sector.

Before the zones were locked and sealed the missing documents had to be located and brought under American control.

By a process of detective work and, where that failed, by persuasion and coercion of the German scientists and programme managers, boxes of blueprints and notes were pinpointed, then recovered from the bottom of lakes, as well as caves, farms, crypts, hospitals and mines.

O'Brien's team hit pay-dirt early: their sleuthing led them from Thumersbach to an air raid shelter covered with earth in the side of a mountain close to the Austrian border. There they found files belonging to the 1 Gruppe/6 Abteilung, the German Air Ministry's intelligence directorate. These detailed all the Luftwaffe's latest air weapons, from the Me 262 jet fighter and the Me 163 rocket fighter to radars, air-to-air missiles and guided bombs. It also showed that the blueprints had recently been smuggled out of Germany in U-boat shipments to Japan.

Reading through the signals transmitted by *Lusty* field team officers to their superiors, I was hit by the sheer scale of the operation and the pressures endured by those who took part in it.

On 22 April 1945, two and a half weeks before the guns in Europe fell silent, additional recruits were needed to process the data. 'It is planned to expand the activities of air technical intelligence ten-fold, securing the most highly qualified specialist personnel available to the Army Air Forces,' Brigadier General George C. McDonald, head of USAAF Intelligence, wrote that day.

The calibre of the technological spoils was indicated by the arrival in-theatre at the end of that month of a 'special group of scientists' headed by Dr Theodore von Karman, special consultant to the US Army Air Forces' supreme command.

The hardware that awaited them was detailed on several pages of microfilmed documentation from McDonald's intelligence directorate. Among them were a jet-propelled helicopter 'in flyable condition accompanied by a complete set of documents and detailed drawings'; the Lippisch P-16 tailless rocket-propelled research aircraft whose advanced construction indicated 'possible operation at high Mach numbers in the vicinity of 1.85'; and the Horten Ho 229 twin-jet flying wing bomber.

Nothing comparable existed in the American inventory; or anywhere else, for that matter.

But by midday, I had still found nothing that indicated the merest presence of the machines elucidated within the Schriever/Vesco legend.

As more men and resources were ploughed into *Lusty*, the search for high technology widened. The files detailed the growing number of shipments back to the US. 'Hanau: 50 tons of documents ready for shipment. Munich: 30 tons of documents ready for shipment. Teisendorf: Approximately two tons of documents pertaining to flight-path studies of guided rocket missiles are on location ready for shipment . . .' From all over Germany, documentation was being gathered, marshalled and hauled back to the US.

But just when it seemed as if the USAAF mission was breaking the back of the task, new finds and new problems added to the logjam. The biggest and most unexpected of these related to the Nazis' underground facilities, whose scale and number had not been anticipated by the investigators, because Allied reconnaissance had failed to pick up anything but the vaguest hints that the Germans were putting their factories underground.

The underground facilities were of more than passing interest to me since they figured in the Legend as places where the Nazis had developed their anti-gravity technology.

From the message traffic that built up around the discovery of these facilities in the late summer of 1945, three or four months after the war in Europe ended, it was clear that the discovery of these underground facilities was diverting the *Lusty* investigators from their primary goal of plundering German technology.

On 29 August, Gen McDonald sent USAAF headquarters in Europe a list of six underground factories that had been discovered and 'excavated'. All of them had been churning out aircraft components or other specialised equipment for the Luftwaffe until the very last day of the war.

According to McDonald, the facilities varied in size from 5 to 26 kilometres in length, according to measurements of their tunnels and galleries. The dimensions of the tunnels varied from 4 to 20 metres in width and 5 to 15 metres in height; the floor space from 25,000 to 130,000 square metres.

Seven weeks later, in mid-October, in a *Preliminary Report on Underground Factories and Facilities in Germany and Austria*, senior USAAF officers were told that the final tally showed 'a considerably larger number of German underground factories than had hitherto been suspected'.

In addition to Germany and Austria, the underground building pro-gramme had been extended across France, Italy, Hungary and Czechoslovakia.

'Although the Germans did not go underground on a large scale until March 1944, they managed to get approximately 143 underground factories into production by the last few months of the war,' the report stated. A further 107 facilities had been located that were either being built, excavated or planned by the end of hostilities, but another 600 sites could be added to the total if caves and mines, many of which had been turned into production lines and weapons laboratories, were taken into consideration.

The report's author was evidently taken aback by the breadth of the German underground plan. 'It is a matter of conjecture what would have occurred if the Germans had gone underground before the beginning of the war,' he concluded.

And then, hot on the heels of the underground investigation, a fresh directive, this time from a senior USAAF field officer to General McDonald, the air force intelligence chief at Wright Field in Ohio.

Set against the steady but predictable build-up of documentation, its tone was so unexpected and the content of the message so different that it took me a moment to come to terms with its implications. It was dated the 28th September, 1945:

1. It is considered that the following have been thoroughly investigated and have proven to have no basis of fact.
  a. Remote Interference with Aircraft
Investigations have been completed on this subject and it is considered that there is no means presently known which was in development or use by the German Air Force which could interfere with the engines of aircraft in flight. All information available through interrogations, equipment and documents has been thoroughly investigated and this subject may be closed with negative result.
  b. Balls of Fire
As far as can be determined from extensive interrogations, investi-gation of documents, and field trips, there is no basis of fact in the reports made by aircrews concerning balls of fire other than that

phenomena similar to balls of fire may have been produced by jet aircraft or missiles. This subject may be considered closed with negative results.

A.R. Sullivan Jr.,
Lt. Colonel, Sig C.

This was the first hint or mention of anything connected with the foo-fighters. Sullivan was telling McDonald that investigations on the subject had been completed; and yet, there was no sign in any of the raw communications traffic that had hitherto flowed between the front and Wright Field that anyone was the slightest bit interested in the subject of 'remote interference with aircraft' or 'balls of fire'. Furthermore, there wasn't any suggestion of interrogations having taken place or equipment or documents having been investigated pertaining to possible foo-fighter technologies – or of special 'field trips' having been undertaken to track them down. But the instruction that the subject 'may be considered closed with negative result' was so strident it made me sit up and wonder if I'd missed something.

I felt a knot of excitement in the pit of my stomach. The subject, as Sullivan put it, *may be closed with negative result*.

Why?

In the apparent informality of its memo format, the implication hadn't been clear on a first read-through. But it was now. This wasn't a casual instruction. It was an order.

I flicked back a few frames and found a field report of an investigation into German guided weapons technology.

It was then that my eyes were drawn to an assessment of a sensor called *Windhund* ('wind-hound'). *Windhund* was a 'sniffer' device that detected the presence of aircraft by measuring differences in the polarity of the surrounding air. It then automatically directed its parent aircraft to follow the trail until it reached the bomber stream itself.

Christ, all you had to do was hard-wire *Windhund* into an autopilot and you had a mechanism to allow a pilotless aircraft – conveivably, the Fireball – to be controlled automatically.

In the same document, a second sensor from a related facility (same organisation, different location) was detailed as an infra-red tracking system for locating aircraft exhaust-gas. Both technologies were at an early stage, but the fact that they were being investigated at all was significant. A half century later, as far as I knew, neither had been perfected or developed. But here, in 1945, were sensors that a foo-fighter could have used to get close enough to a bomber to disrupt its engine ignition systems.

I felt for the first time like I had a foot in the door.

A few frames later and here was a large document set out in tabular format with headings running across the top of the page: *Target, Organisation, City, Activity, Assessed (date), Action Taken and Remarks.* From the first page, it was clear that this was a very different kind of file, filled with data as raw as it came, and instantly noteworthy for what it did – and didn't – say. In the remarks section of one target, listed only as a 'research station' near the town of Eib See, assessed primarily by the British, curiously, on the 2nd May 1945, it said: 'Over 400 evacuated Peenemunde personnel held. Excavations made in the mountainside close to lake for underground workshops. A very important target.' Nothing else.

Mysterious as hell.

Next page.

Target: Luftfahrt forschungs [sic]. *City*: Brunswick. *Activity*: Radio-controlled aircraft. *Assessed*: 21–24 April, 1945. *Action Taken*: Team has been dispatched. *Remarks*: 'Evidence of radio-controlled aircraft.'

I scratched some notes. Another potential foo-fighter technology.

Next page.

Target: Research. *Location*: 87, Weimarerstr. Vienna. *Activity*: Experiments with anti-aircraft rays. *Remarks*: 'Research activity is conducted in a house at the above address. Research personnel were not allowed to leave house (reported hermetically sealed).'

Jesus.

The Germans *had* been working on directed energy weapons and had sealed those working on it from the outside world.

So many radical technologies and so long ago. I had no idea.

Next page.

Target: Daimler-Benz. *City*: Unter-Turkheim/Stuttgart. *Activity*: Secret weapon. *Assessed*: 25 April 1945. *Remarks*: 'Said to stop ignition system of a petrol engine. The apparatus has succeeded in stopping a motor vehicle w/magneto ignition, but not one w/battery, at a range of 2 or 3 km.'

The report went on to say that the technology had been insufficiently developed to have been brought against aircraft, but it made clear that this was the goal.

The Germans, then, *had also* been working on devices that were designed to 'interfere with the engines of aircraft in flight', contrary to the note sent by Col Sullivan to Gen McDonald on the 28th September.

Why, then, did Sullivan say that all information available through interrogations, equipment and documents had been 'thoroughly investigated' and that the subject 'may be closed with negative result'?

Target	Location	Source	Activity	Remarks
Henschel Factory	Oderwerke (Nordhausen)	OSS London - XL 2285, 13 April 1945.	Manufacturing parts for the HS-177 Rocket AA Weapon.	Mfg. actually takes place underground in an old mine shaft.
Henschel Factory	Hötzleben (Nähle-Kässel)	ditto	Manufacturing parts for the HS-177 Rocket AA Weapon.	
Laboratory	Grosser Inselsberg (Northern end at the Thuringen Wald)	A.D.I.(K) Rpt. 34/1945.	Proximity fuses (electrostatic)	
Friedrich Listwerke ?/m	Plauen - Vogtland	OSS-London, XL-2285, 13 April 1945.	Mfg. remote control mech. ("Kreisel") for the projectile of HS-177 Rocket (AA Weapon).	
Research	87, Weimerstr. Vienna	OSS-London No. B-2222 (23) 13 April 1945.	Experiments with anti-aircraft rkts.	Research activity is conducted in a house at the above address. Research personnel were not allowed to leave house (virtually hermetically sealed)
Research	Bheide ? ???	P.W. Report by A.D.I.(K) No. 228/1945.	"Versuchsgebuiss" consisting partly of a Kinetheodolite posts for observing the flight of rockets launched from Peenemünde.	
School	Lubeln	A.D.I. (K) No. 226/1945.	Instruction on the beam system used in connection with A-4 Rocket.	Students received instruction on the "Leitstrahl" (beam) system. This beam system used in connection with A-4 rocket. The P.W. in this report confirms the A.D.I.(K) Rpt. No. 632A3984 that one of 3 batteries of an obtaining was equipped with the Leitstrahl transmitter.

What negative result?

Over the years, I had conducted extensive interviews on and off the record with intelligence analysts on both sides of the Atlantic about Soviet and Russian weapons developments and knew something of their techniques. They would never have dismissed an invention such as *Windhund* as cursorily as Sullivan had in the light of clear evidence that the enemy had been developing such a radical strand of technology. At the very least, the reports would have gone into a holding file. If no other reports came in to substantiate this finding, then in time it would have been downgraded. Only then, would field agents have been told to stop looking.

Instead, Sullivan's exhortation for investigators to cease forthwith their search for one of the key component technologies of the foo-fighter sightings, a mystery that had absorbed the USAAF's intelligence community in that winter of 1944–45, was, well, odd.

Unless, that is, they had found a foo-fighter and needed to put the lid back on.

It got better. I came to the field report on the last target to have been assessed, the Institut für Elektrophysik Hermann Goering at Landsberg-am-Lech in Bavaria. Like so many of the others it was quite anodyne until the remarks section, which read:

'Experimental work in conjunction with airfield at Pensing, 9 km north (toward Munich). Activities: research on gas, aircraft, pilotless aircraft, radio-communication.'

Here, then, was a facility that merged in a single programme all the activities that explained Vesco's Fireball drone: a pilotless, remote-controlled aircraft that disrupted the engines and electronic systems of Allied bombers. And the fact that the research activity was taking place at an airfield seemed to suggest that the tests were at an advanced stage when the *Lusty* assessment team showed up.

I sat staring at the screen for several more minutes, scouring the text for other corroborative details, but there were no further clues and I was out of time.

I photocopied what I could, thanked the archivist and made my way outside. The taxi was already waiting, engine running, wipers working against the rain. I threw my bags on the back seat and sat back for the ride, my head buzzing with thoughts about a strand of weapons technology that appeared to have been buried from view for more than half a century.

But it wasn't until the flight home that sanity prevailed.

What had I actually learned from *Lusty*?

That the Germans had been working on technologies that far exceeded published accounts of their aeronautical achievements in the Second World War?

Certainly.

That they had developed a craft that explained the foo-fighter sightings reported by Allied airmen?

Possibly.

That they had developed anti-gravity technology – and that this technology had fallen into the hands of the Americans at the end of the war in Europe?

Unfortunately, not.

Besides which, I realised that *Lusty* had been declassified in the mid-1960s, several years before Vesco's book *Intercept, But Don't Shoot* was published. A sceptic could claim that Vesco had simply concocted his account from the documents.

I had hoped that by applying a few basic investigative principles, I would find some tangible evidence of a radical propulsion technology that had been overlooked by everyone else.

Of course, it was never going to be that simple.

Despite all the stories, there was no concrete proof that the Germans had even designed, let alone built, a flying saucer. As for anti-gravity propulsion, no one had ever found a single piece of evidence that would reliably hold up in court.

A short post-script to the German story emerged; in Canada of all places. At the time it seemed irrelevant to the main body of my investigation, but it came screaming back at me later in a way I couldn't possibly have guessed at so soon after my return from Washington. I would recall it later as the 'Silverbug affair'.

In April 1959, during hearings of the Space Committee of the US Congress, the US Department of Defense revealed that it was working with the Avro Canada company to develop a 'flying saucer'. According to the Pentagon's Research and Engineering Deputy Director, John Macauley, the craft would 'skim the earth's surface or fly at altitudes reached by conventional aircraft'. This was the first public admission that Avro was building a flying disc, the Avrocar, the brainchild of a gifted British engineer called John Frost, who'd moved to Canada shortly after the war.

Activity surrounding the Avro project was shrouded in mystery. Word of its existence had leaked as early as 1953, when the *Toronto Star* reported that the company was working on a 'flying saucer' that could take off and land vertically and would fly at 1,500 mph. British and

American scientists were also reported to be involved. Both the Canadian government and Avro played down the reports.

When the Avrocar finally emerged from the shadows it was easy to see why. One look at the prototype was enough to know that it would never achieve supersonic flight. Flight tests, in fact, confirmed it was a dog.

Underpowered and unstable, the Avrocar made its first untethered flight in December 1959 and soldiered on for two more years as Frost and his team sought to overcome its failings.

By the end of 1961, at the conclusion of the US development contract, the Avrocar had managed to claw no more than a few feet into the air. Funding for the world's only known flying saucer programme was not renewed and the Avrocar drifted into obscurity – although not fast enough for the Canadian aerospace industry, which has sought ever since to distance itself from the project. There and elsewhere, it has become a byword for failure; something of an aviation joke.

Yet, that may have been the intention all along.

Recently declassified papers show that from early 1952 right up until the cancellation of the Avrocar in 1961, a select group of engineers known as the Avro 'Special Projects Group', led by John Frost, had been working on a highly classified set of programmes that accurately reflected the *Toronto Star*'s original leak of 1953.

The papers revealed that, commencing with Project Y, also known as the 'Manta', via Project 1794 to Project PV 704, the SPG tested technology for a whole family of flying saucers whose projected performance was designed to eclipse that of all other jet fighters. Data on Project 1794, a design for a perfectly circular fighter-interceptor, show that it would have been capable of Mach 4 at 100,000 ft. The best fighters then on the drawing board were hard-pressed to achieve Mach 2.

On 2 December 1954, Canada's Trade Minister C.D. Howe sought to dampen speculation about the existence of an Avro flying saucer programme with a statement that the Canadian government had backed studies into disc-shaped aircraft in 1952–53, but had since dropped the idea.

This, as it turned out, was true – up to a point. In fact, the Avro design, now called Project Y2, had been purchased outright by the US Air Force and recodenamed 'Project Silverbug'.

Under pressure from the US press to reveal the nature of this programme – which, though classified, had again leaked – the USAF issued a bland statement in 1955 that it had a 'research and development contract with the Avro Company of Canada to explore a new aircraft design concept.' But that was all.

USAF documents on Silverbug declassified as recently as the mid-

1990s show just how radical the Silverbug/Y2 concept was. 'The Project Y2 design proposal incorporates a number of advanced improvements brought about by the utilization of several radical ideas in fundamental areas.' Among these was a 'very large radial-flow gas-turbine engine, which, when covered, will form a flying wing with a circular planform, similar in appearance to a very large discus.'

The entire air vehicle, in other words, was a giant flattened jet-engine, the hot gases that propelled it forward exiting from slots around its rim. By directing the exhaust, Silverbug would have been capable of almost instantaneous high-speed turns in any direction; manoeuvres that would have included 180-degree reversals, the craft flipping this way and that like a tumbling coin, and all with minimum discomfort to the pilot, because the g-forces would have been alleviated by the adjustable thrust.

Much like the craft that had vexed the US Air Force intelligence community following the first rash of US flying saucer sightings in 1947.

The only other place prior to this discovery I had ever heard mention of a radial-flow gas-turbine, an RFGT, was in Germany during the war.

After the ignominy of the Avrocar programme, the Avro SPG disbanded and most of its exceptional brain-talent was snapped up by US aerospace companies or government-run research establishments. The most notable exception was John Frost, the maverick genius and inspiration behind the saucer projects, who soon after the Avro company was dissolved in April 1962 emigrated to Auckland, New Zealand. After the exhilaration and excitement of designing highly classified, highly unconventional air vehicles for the Canadian and US air forces, Frost settled for the relative obscurity of an engineering job with Air New Zealand. He rarely, if ever, spoke about his work on the Avrocar and never mentioned the existence of the supersonic saucer projects that had gone on all the while behind it. He died of a heart-attack in October 1979 at the age of 63.

Soon after I got back from Washington I rang Frost's son, Tony. I had tracked him down to Auckland, New Zealand, via a small band of researchers, all with strong aerospace credentials, who had done a fantastic job in piecing together the Avro story – the real story; one in which high-performance flying saucers had been shown to be real; far from the stuff of legend, as they had been in Germany.

'Dad was very secretive about his work,' Tony Frost told me. 'Everything I ever learned about what he had been doing in Canada I had to pick up from people he'd worked with. The Special Projects Group was, for a time, just about the most secret group of its kind in Canada or the US, but why its activities should remain shrouded in secrecy for so long is still a mystery to me.'

In the brief silence that followed, my mind worked hard to fill in the gaps. Had the lumbering Avrocar been an intentional blind all along, something to draw the flak from the SPG's real, but hidden goal: the creation of a supersonic, vertical take-off and landing interceptor?

With Avro shut down, had work on the supersonic saucers continued in the utmost secrecy in the US? Was there something about the very shape and form of the saucer, indeed, that was inherently super-classified?

While the Avro programme didn't entirely explain how saucer-shaped craft had been 'within the present US knowledge' in 1947, it came very close.

First-off, Avro showed that man-made flying saucers were real; they'd existed.

Secondly, if you substituted the radial-flow gas-turbine, the RFGT, for something even more exotic – a T.T. Brown-type propulsion system, for example – the specification was a pretty neat match.

In the conventional world of aerospace, testing a new kind of airframe with a known engine was standard operating procedure. Was this what the Avro programme had really been about? Were the Americans using the Canadians to perfect the aerodynamics of the saucer before the arrival of an all-new propulsion-source – something that Trimble and Co had hinted at in their statements of 1956?

Or was the truth altogether more prosaic? Had the supersonic Avro disc programme merely been hushed up all these years in an attempt to avoid embarrassment by those who had thrown money at it and seen it fail?

These were questions without answers, so I asked Tony Frost instead how he thought his father had come by his radical ideas.

He told me, almost in passing, that a file recently uncovered in Canada's National Archives had shown that his father had made a journey to West Germany in 1953. There, at a Canadian government installation in the company of British and Canadian intelligence officials, Frost met with a German aviation engineer who claimed to have worked on a vehicle similar to the disc-shaped aircraft on the drawing boards at Avro. The German said that the project had been under way at a site near Prague, Czechoslovakia, in 1944–45, and that the saucer had not only been built, but flight-tested. He told Frost that both the plans and the craft itself had been destroyed in the closing weeks of the war.

How the information was used by Frost was never made clear. Like the

Avro supersonic disc programme, he never spoke about it – to his family or anyone.

I felt the historical data pulling me back. If the Canadians and the British had been colluding on German-derived saucer technology, it was dangerously close to the scenario that Vesco had espoused all along.

But Tony Frost had spent years trying to piece together the secret history of his father's life and he was still no closer to the truth.

I had exhausted the safe, cosy world of archives and Internet trails. It was like Cross had said; they'd succeeded only in taking me everywhere and nowhere. If a decade of training had taught me anything, I knew my only real chance of verifying Trimble's outburst back in 1956 was to search for leads in the real world.

I didn't know it yet, but I was about to get a helping hand from a scientist on the inside loop, a man I came in time to regard as a genius, who for reasons of his own had already embarked on exactly the same journey as me.

Chapter 7

I had expected the drive to Austin, the state capital, to provide me with a rich set of impressions of the frontier that divided the Eastern Seaboard from the wilderness that started here and extended to the Pacific. But for two hours I sat on the freeway, my views of Texas obscured by bumper-to-bumper traffic, peeling trailer-homes and second-hand car lots. And so I thought of the man I was about to visit instead.

Dr Harold E. Puthoff, 'Hal' to those who knew him better than me, was the director of the Institute of Advanced Studies at Austin. Puthoff had lived his whole life on a frontier; one that straddled known science, the world explained and understood, and a place that most people, scientists included, said didn't exist.

I felt drawn to Puthoff, because his academic qualifications, not to mention his connections to the military, put him squarely in a place that I could relate to; yet, he was clearly also involved in some shit of the weirdest kind.

His résumé, which I had pulled off the Internet, said that he had graduated from Stanford University in 1967 and that his professional background had spanned more than three decades of research at General Electric, Sperry, the Stanford Research Institute (SRI) and the National Security Agency, the NSA, the government's super-clandestine electronic eavesdropping organisation.

It also told how he 'regularly served various corporations, government agencies, the Executive Branch and Congress as a consultant on leading-edge technologies and future technology trends' and that he had patents issued in the laser, communications and energy fields. Puthoff was no lightweight.

What his résumé didn't describe was one of the strangest episodes in the history of US intelligence-gathering: a programme, known as remote viewing or RV, that enabled the Central Intelligence Agency and the Defense Intelligence Agency to spy on America's enemies using clairvoyance. For years, depending on whom you believed, this highly organised, well-funded operation had proven remarkably successful, providing reams of usable data on Soviet military activities that were inaccessible to more traditional intelligence-gatherers, such as spy satellites and aircraft.

RV remained buried, deeply classified, until 1995, when the CIA finally admitted it had used 'psychic spies' against Russia during the Cold War.

Puthoff had been the RV programme's founder and first director and had run it for 13 years, finally leaving SRI in 1985 to establish his own institute, the IAS, at Austin.

And now, he was up to his ears in gravity work.

If the US government had instituted a top-secret anti-gravity programme, decades before possibly, then chances were Puthoff was either directly involved, knew about it, or, at the very least, had some inkling of it. On the other hand, doing what he'd done, acting as a consultant to some of the government bodies he still advised, I knew that Puthoff breathed the same air as people for whom disinformation was a just another part of the toolkit.

Ordinarily, I would have been wasting my time and Puthoff's. My limited, workaday knowledge of physics put me at an enormous disadvantage when it came to sorting out good information from bad in the anti-gravity field.

But on that day, as the miles to Austin sped down on the clock, I had some help.

Deep in the heart of Britain's own aerospace and defence establishment, I had found someone to decode the intricate mysteries of gravity for me.

I wound back the clock to the time when I'd attended a trade conference many years earlier. It had been entitled 'Anti-gravity! The End of Aerodynamics'. I'd only gone along because of the calibre of the speaker: Brian Young, Professor of Physics at Salford University. Young was also Director of Strategic Projects for British Aerospace (BAe) Defence, Britain's biggest military contractor.

The lecture had been given at the London headquarters of the Institution of Mechanical Engineers, IMechE, just off Parliament Square. Handouts beforehand (which I'd clipped to my copy of the transcript) had set the theme: 'Gravity is the most mysterious of all the natural forces. It appears completely indifferent to anything we try to do to control it. Nonetheless, its very presence is the driver for the whole science and business of aeronautics. Scientists from the very distinguished to the very ridiculous have tried for over 300 years to either explain it or destroy it.

Technology is making tremendous strides in all directions and perhaps we are getting close. If so, is this the end of aerodynamics?'

I chewed over the words of Young's speech, according it a respect I hadn't given it on the night. Now I wished I had.

I recalled the elegant marbled hallway and the low hubbub of conversation as I'd drifted among the attendees – a hotchpotch of MoD men and aerospace industry executives, some of whom were familiar to me – waiting for the doors to open into the auditorium.

Young had kicked off his talk by reiterating our general state of ignorance about gravity. 'It is an incredibly weak force,' he'd said, 'although it probably didn't seem so to the Wright Brothers and I suspect most mountain-climbers would disagree violently.

'But really gigantic quantities of matter, about 6x10 (to the power of 21) tonnes in the case of the earth, are required to produce the gravity field in which we live. If you think of a simple horseshoe magnet lifting a piece of iron, then, as far as that piece of iron is concerned, the magnet weighing a few ounces is outpulling the whole earth.'

He flashed up a set of artist's impressions showing how three different kinds of anti-gravity vehicles might look assuming engines could be built for them.

One, a 'heavy-lifter', looking much like an airship, was depicted effortlessly transporting a giant section of roadbridge through the air. Young made the assumption that the anti-gravity engine of the heavy-lifter not only cancelled the weight of the craft, but its underslung load as well.

Another, saucer-shaped, craft, described as being like a bus for 'city hopper' applications, used a 'toroidal' or doughnut-shaped engine to allow it to skip between 'stops' in a futuristic-looking metropolis.

The third showed a combat aircraft with a green-glowing anti-gravity lift-engine on its underside for vertical take-off and landing (VTOL) and conventional jet engines for forward propulsion.

But all these things were fiction, he stressed, so the designs were simply conjecture. 'The more I have read and thought about anti-gravity and its terrestrial applications, the more I have become convinced that as it stands today it is going nowhere. Whatever name the press may give experiments, either in someone's garage or in Geneva, most of the real work is trying to uncover basic knowledge and few people really believe it will lead to gravity control.

'To the contrary, a view I have seen expressed is that if gravity control is ever discovered it will be as an unexpected by-product of work in some completely different field.'

To which he'd added a rider. Were he to have got it all wrong, were an anti-gravity effect ever to be discovered, the growth of the industry surrounding it would be exponential – look what had happened after Faraday's discovery of electro-magnetic induction in 1831. Within five years, the electric motor industry was born. Within seventy, the generation of electricity was huge business. Today, it is hard to conceive of a time when man-made electricity did not exist.

I vaguely remembered going up to Young after the lecture and chatting to him about some of the points he'd raised. But now I kicked myself. If only I'd paid more bloody attention.

Soon after I'd returned from the low-key archive in Washington that housed the *Lusty* documents, I'd picked up the phone and dialled British Aerospace. Through to the public relations department and I logged the request. Could they get me Professor Young's phone number?

Forty minutes later, the media manager called back to say that Professor Young had died, of a heart-attack he thought, some years earlier.

I put the phone down, a little shocked at the news, and pinched the bridge of my nose, looking for that extra bit of concentration. What had Young and I talked about? Down into the basement and another protracted rifle through a box taped up and marked 'notebooks: 1989–94'. There in amongst my old reporter's pads, after an hour on my hands and knees, I found the one that I'd taken with me that night to IMechE. I started reading and flicking pages.

The notes I'd jotted down were piss-poor, highlighting terms like 'gravito magnetic permeability' and 'toroidal engine' and putting big question marks against them.

I kept turning. There, right at the end, I'd written myself an aide-memoire to put in a call to a media affairs officer at BAe who would arrange for me to receive photos of the artist's impressions Young had flashed up during his speech.

This, I remembered now, was one of the things Young and I had discussed.

Below the aide-memoire were two names: Dr Ron Evans of 'BAe Defence Military Aircraft's Exploratory Studies group' and Dr Dan Marckus, an eminent scientist attached to the physics department of one of Britain's best-known universities. The department was more than familiar to me. Its work was well-known in defence circles for the advice that it gave to the government.

Beside Evans' and Marckus' names were their contact phone numbers. I called Dr Evans first.

He turned out to be a nice, soft-spoken mathematician based at BAe's Warton plant, a facility that cranked out jet fighters close to where the bleak mudflats of Lancashire's Ribble Estuary met the Irish Sea.

When I asked if he could brief me, as per the suggestion of Professor Young, Evans said he needed to consult with the company's media managers. He'd never dealt with the press before. I got the impression it was sensitive stuff.

It was, but not for any of the reasons I had guessed or hoped. Britain's biggest aircraft manufacturer, BAe, had an anti-gravity department. That in itself was key.

But its activities weren't classified. Far from it. The impression I got, more than anything else, was that the management had tucked Evans' tiny department into the farthest-flung corner of the company because it didn't know what else to do with it. That and the fact no one in BAe at that level wanted to wake up to headlines that it was involved in kooky, Buck Rogers science. It would destabilise the share-price.

In the end, Evans wasn't allowed formally to brief me on his work, but he was given permission to fill me in on some of the background that had led to the formation of his Exploratory Studies group. That suited me fine. BAe didn't want to break cover – nor did I. For once, we were on the same side of the secrecy divide.

In March 1990, Evans chaired a two-day University–Industry Conference of Gravitational Research, sitting around a table with a gathering of distinguished academics to identify any emerging 'quantum leaps' that might impact on BAe's military aircraft work. Gravity-control figured extensively on the agenda.

Imagine it. A technology, popping out of nowhere, that rendered all of its current multi-billion-pound work on airliners and jet fighters redundant at a stroke. Now that, I thought, *really* would do things to the share-price.

The company also undertook some practical laboratory work in a bid to investigate the properties of a so-called 'inertial-thrust machine' developed by a Scottish inventor called Sandy Kidd.

In 1984, after three years' work building his device – essentially, a pair of gyroscopes at each end of a flexible cross-arm – Kidd apparently turned it on and watched, startled, as it proceeded to levitate, then settle three inches above the surface of his work-bench.

An ex-RAF radar technician, the Scotsman had become obsessed with the idea of inventing an anti-gravity machine after he'd removed a still-spinning gyroscope from a Vulcan nuclear bomber. Carrying it backwards down the aircraft's ladder, he reached the ground and was promptly flattened by a sledge-hammer blow to his back. The force, he realised, had been transmitted from the gyroscope, which had reacted in some inexplicable way as his feet had touched the ground.

In May 1990, BAe began a series of trials to test whether there was anything in Kidd's claims, knowing full well that he wasn't alone in making them.

In the mid-1970s, Eric Laithwaite, Emeritus Professor of Heavy Electrical Engineering at Imperial College London, demonstrated the apparent weight loss of a pair of heavy gyroscopes by lifting the whole whirling contraption with one hand and wheeling it freely around his head. When the twin rotors were not in motion, attempting this feat was impossible. The gyroscopes were the size and weight of dumb-bells. But when the rotors turned, they became as light as a feather.

Somewhere along the line, the machine had lost weight.

The accepted laws of physics said that this was not possible, out of the question – heresy, in fact. But Laithwaite's claims were supported by a top-level study into gyroscopes published by NATO's Advisory Group for Aerospace Research and Development (AGARD) in March 1990.

The authors of the AGARD report concluded that a 'force-generating device' such as Laithwaite's, if integrated into a vehicle of some kind, could, in theory, counteract gravity. 'Clearly if such a counteracting force was of sufficient magnitude it would propel the vehicle continuously in a straight line in opposition to said field of force and would constitute an anti-gravity device.'

The report went on to say that there was at least one 'gyroscopic propulsive device' that was known to work and that the inventor, E.J.C. Rickman, had taken out a British patent on it. The trouble was, the report concluded, the impulses generated by these machines were so slight they would be useless for all practical applications – except, perhaps, to inch a satellite into a new orbit once it had already been placed in space by a rocket.

It was hardly a quantum technological leap. But that wasn't the point, Dr Evans told me. What was being talked about here was an apparent contravention of the laws of physics; the negation, at a stroke, of Newton's Third Law, of action-reaction. Which was why the BAe-sponsored tests on the Kidd machine had a relevance that went way beyond their immediate and apparent value.

If there were ways of generating internal, unidirectional, reactionless forces in a spacecraft, and in time they could be refined, honed and developed, the propulsion possibilities would be limitless.

In May 1990, a trial went ahead under the watchful eyes of two wind-tunnel engineers from BAe's Warton aircraft plant. The first series of tests showed no anti-gravity effects, but to everyone's surprise, during a second series of trials, one result did appear to indicate that the machine had changed weight.

Unfortunately, the engineers failed to repeat the result; and in science and engineering, repeatability is the benchmark of success.

Had it been repeatable, the two per cent weight change Kidd claimed for his device would have been scientifically verified, and BAe would have employed the world's first known anti-gravity device as an orbital manoeuvring motor for satellites.

Shortly after the trial, Cold War defence budgets collapsed and BAe's share-price went into free-fall. As it fought for its survival, there were cut-backs across the company and Dr Evans' department, with little hope of any short-term returns, was disbanded as quietly as it had come together.

That, essentially, was how things were when he and I met.

I rang Dr Marckus a few times, but only ever got an answer-machine and so I capped my involvement in the BAe anti-gravity story with a few low-key articles in *Jane's* about the BAe investigation. It was more an excuse than anything else to broadcast a question: was anyone else out there engaged in similar work?

The US Air Force had recently proclaimed its interest in the field with a document, published in August 1990, called the *Electric Propulsion Study*. Its objective was to 'outline physical methods to test theories of inductive coupling between electromagnetic and gravitational forces to determine the feasibility of such methods as they apply to space propulsion.' Stripped of the gobbledygook, it was really asking whether there was any theory out there that might permit the engineering of an anti-gravity device.

The existence of the *Electric Propulsion Study* gave me an excuse to slip tailored messages into a couple more stories about futuristic propulsion. It felt a little like pointing a transmitter into deep-space and waiting for a response.

I waited, but nothing came back. No emails, no anonymous faxes, no spooky phone-calls. Nothing.

Thereafter, I kept a weather-eye on the anti-gravity scene, but the receiver I had rigged for the slightest trace of life out there remained silent. In the end, I switched it off.

During the mid-1990s, BAe kept its toes in the gravity field by placing a few small-scale contracts with a number of British universities, principally to see if there might be a provable relationship between gravity and heat; a link first postulated by Michael Faraday in 1849.

Like me, the company was clearly reaching, but getting nowhere.

Then, in 1996, NASA – the US National Aeronautics and Space Administration – began questioning some of the perceived limits to the advancement of air and space travel in two notable areas. One related to the unbreachability of the light-speed barrier – a throwback to Einstein, whose Theory of Relativity said that the velocity of light was a fixed constant never to be exceeded.

The other questioned the heretical notion that, somewhere in the universe, there might be an anti-gravity force; one that could be exploited for air and space travel.

The resultant NASA programme, wrapped within something called ASTP – the Advanced Space Transportation Programme – prompted BAe to resurrect its anti-gravity work. It has since been rebranded 'Project Greenglow', a 'speculative research programme ... the beginning of an adventure which other enthusiastic scientists might like to join, particularly those who believe that the gravitational field is not restricted to passivity.'

Decoded, Greenglow involved little more than re-seeding small amounts of company R&D money to more British universities, funding a few more experiments of the Kidd kind, and acting as a European feed-in to the NASA effort.

It was, to say the least, depressing news. I thought back to George S.Trimble, the man who had been too afraid to talk to me after Dani Abelman of Lockheed Martin had approached him on my behalf. If Trimble had cracked the gravity code back in 1956, why were BAe and NASA, with their immense know-how and resources, still struggling with the gravity conundrum today? It didn't make any sense.

And that should have been where the matter ended.

But long months after I had consigned the anti-gravity story to the graveyard, I got a phone-call. The man at the other end of the line was bluff and unapologetic, verging on the rude. Dr Marckus had finally got in touch.

Chapter 8

The sunlight reflected brightly off the thin carpet of snow that had fallen along the east coast overnight. I kept the estuary on my right and the sea defences on my left, edging the car along the narrow road, navigating by the instructions I had been given over the phone. There was something almost primeval about the combination of the pine forest, the limbs of the trees twisted into impossible shapes by the wind, the open sky and the stillness.

The trees parted briefly and I saw the radar mast that I'd been told to look for. Around another bend and the tiny haven swung into view.

I pulled up in the car park beside a wooden jetty that jutted into the green, fast-flowing current towards a small settlement on the far bank. The buildings seemed run-down and dirty compared to the crisp scenery through which I'd just driven.

There were two other cars parked beside mine and both were empty. I got out, stretched my legs and shivered as the still-freezing temperature blew away the fug of the hour-and-a-half-long journey from London. Suffolk. The East Anglian coast and the cold waters of the North Sea. I was in a part of the country I didn't know.

A light breeze tugged at the rigging of a row of yachts dragged onto a thin stretch of stony beach nearby. Seabirds picked along the jetsam and detritus of the shoreline. Above the gentle pinging of the mast ropes, I heard children's voices.

There were four people on the jetty.

A woman sat on the far side, wrapped warmly against the cold, reading a book as her two small children line-fished for crabs.

The father, standing further off, his back to me, was tracking an open motor-boat chugging hard against the fast-flowing water, its helmsman preparing to tie up alongside the jetty's end.

As the boat bumped against the tyre buffers, the man stepped on board and settled onto one of the bench-seats.

I waited for the woman and the two children to get up and join him, but they never so much as glanced his way.

It was then that I realised the man in the boat was Marckus.

He looked up and waved me aboard. I hesitated. Taking a ride on a boat had not figured in our brief phone conversation.

Until then, I'd not considered the possibility that Marckus was anything other than the person he claimed to be.

But what did I know about him? That he was plugged firmly into Britain's aerospace and defence community, that he didn't make a habit of talking to journalists, and that he was taking a risk by talking to me.

The ferry was preparing to leave.

I ran along the jetty and dropped down into the boat. Marckus paid for both of us and the ferryman cast off.

I found myself apologising for the mix-up; how I'd thought Marckus was with the woman and her children on the jetty. Silly, nervous jabbering.

Dr Marckus cut me short. 'Does the significance of this site mean anything to you?' he asked, his arm almost hitting my face as he waved in the direction of the road.

I shrugged and he gave a mild snort of derision. 'I thought you said you were a Jane's man.'

'I'm sorry,' I said, 'it looks like a small yacht club and a fishing village.' I neglected to tell him, because I couldn't see the point, that history wasn't my bag. My remit was high-technology of the 21st-century kind.

Once, Dan Marckus must have been lean and ascetic-looking; the thinning hair and high forehead still gave a hint of it. With age, however, he had put on some padding, making him look a lot less severe than he sounded. A dark, scratchy beard, flecked with grey, distracted the eye on first meeting from a thick pair of lenses in tortoise-shell frames.

Behind them, Marckus' brown, beady eyes watched me intently and with a hint of irritation.

He was dressed sharply in a thick blue polo-neck, a trendier version of the kind fishermen wear, and a brown leather jacket. His voice said he was around 65, but he looked younger.

'What you saw when you drove through those trees was one of the orginal antenna masts of Bawdsey Manor. Before the Second World War broke out, the Germans sent an airship, the *Graf Zeppelin*, up and down this coastline trying to figure out what we were up to – what those tall steel lattice towers meant.'

'And what did they learn?' I asked.

'They'd packed the *Graf Zeppelin* with electronics – listening gear. But the old man was so much smarter than them, thank God.'

Marckus registered my bewilderment. 'Robert Watson-Watt. The man who gave the British the secret of radar. It all happened here. I thought you'd know that.'

Once on the other side, we jumped down onto the beach and strolled along the shoreline, past a lonely cafe, in the direction of the point.

'So how did Watson-Watt outwit the Germans?' I asked, to fill the silence more than anything else.

'They saw the *Graf Zeppelin* on their scopes when it was way out over the North Sea, realised what it was up to and switched off the radars. The Germans went away believing that they were radio masts, that Germany was the only nation that had successfully developed radar detection. A month later, they invaded Poland.'

'That's interesting. Did you work for him?' The term he'd used, 'old man', denoted more than a passing familiarity. 'Watson-Watt, I mean.'

Marckus never answered. He walked on in silence, watching the gulls soaring on the cold currents above our heads. Then, without looking at me, he said: 'It's been a while since you've written about my favourite subject. I was beginning to worry that you'd lost interest in it.'

'It's like Professor Young said in his speech. I have come to the conclusion – a little late in the day, perhaps – that anti-gravity is for the birds.'

'I see,' he said. 'Are you hungry?'

I looked at my watch. It was close to midday and I hadn't had any breakfast.

Marckus, registering my hesitation, doubled back towards the cafe.

I pulled a copy of the US Air Force's *Electric Propulsion Study* from my coat pocket and set it down on the table next to a tatty yellowing manila folder that had been clamped under his arm throughout.

Marckus and I were the only two people this side of the deep-frier. After serving us our plates of cod and chips, the cafe's proprietor left us on our own.

Marckus seemed to start to relax.

'Apart from the fact that the maths and the physics are beyond me, the evidence says that this whole science, if you can call it that, is still in its infancy,' I said. 'One look at this report says they're still playing around with algebra. When I wrote what I wrote, I thought the story was bigger than that. I have to concede I was wrong.'

'What were you hoping for, if I may ask?' Marckus asked.

I shrugged. 'I thought maybe they'd have at least built some hardware by now. God knows they've had time. The US aerospace industry first started talking about this stuff in the mid-1950s. You'd think by now there might just be something to show for it. But the Air Force is just staring at its navel and doing a few sums.'

I tapped the *Electric Propulsion* report. 'Modifications to Maxwell's equations and all that.'

'As I recall,' began Marckus, 'in the 1970s, the Americans tripped over some research published by a Russian mathematician, Ufimtsev, who in turn had gained his inspiration from work done by Maxwell a hundred years earlier. The end result was stealth, the biggest breakthrough in military aircraft design since the jet engine. Not bad, for a mathematician.'

'The difference is, they developed stealth,' I told him.

'How do you know they haven't developed anti-gravity?'

I studied his face. How much did Marckus really know?

He must have read my expression. 'My information isn't any better than yours,' he said, raising a hand in the air to signal his innocence. 'The US and the UK share a lot of sensitive data, as you know. But there's nothing come across my desk that says we're cooperating with the Americans on anti-gravity – or even that the Americans are doing it on their own. My point is, we don't know.' He brought his hand down on the manila folder to reinforce the point.

I looked at the US Air Force study again. It was hardly reassuring. If anti-gravity existed and people had known about it years ago, how come the Air Force – along with BAe and NASA – was only just starting to look for it in the mid-1990s?

Marckus stared into his cup. 'Did they build the first atomic bomb with a few "poxy" million bucks? Christ, no. It cost them billions of dollars, even in 1940s values. Secret dollars nobody outside the secret was supposed to even guess at.'

'So, has anti-gravity been developed in the black or not?'

Marckus glanced up. His voice hardened. 'I already told you. I don't know. None of us this side of the fucking Atlantic knows.' Then he coughed, semi-apologetically. 'But there are signs.'

'What signs?' Despite myself, I leant forward.

'Up until 1939, atomic energy was still a matter of conjecture, even amongst physicists in the field. But by the following year, some of the brighter sparks in the community realised that it was doable, that if you split a neutron into equal parts, you'd quite likely get an enormous release of energy. Things that appear impossible usually aren't, even when the physics *say* they are.'

I knew enough to know that in the 1930s next to nothing was understood about nuclear physics, that by the outbreak of the Second

World War all experiments resulting in nuclear reactions had required more energy than had been released.

As more and more physicists escaped the Nazis, however, and debate in America was stimulated further, the conjecture spilled into the US press and writers began to speculate about what they had heard. A number of sensational articles appeared about the weapons implications of a nuclear reaction that could be compressed into the blink of an eye. The word was, it would result in a massive bomb.

By 1940, when it was understood that uranium bombarded with neutrons fissioned and produced more neutrons and that a multiplying chain-reaction really might occur with huge explosive force, something remarkable happened. The nuclear physics community voluntarily stopped the publication of further articles on fission and related subjects and this, in turn, dampened the media's interest.

By the time America entered the war in December 1941, you couldn't find a mention of fission. It was as if no one had ever been discussing it.

The parallel with what I had read in the US media in the mid-1950s, culminating in all those statements in 1956 that anti-gravity was doable with a Manhattan-style effort behind it, and then the crushing silence, was extraordinary.

By 1960, you talked about anti-gravity and people looked at you like you were mad.

I stared out the window, trying to find Watson-Watt's once-secret radar mast, but it was lost behind the trees. I knew Marckus lived somewhere nearby. I wondered where, exactly; what his house was like, whether he was married, if he had kids.

'What's in this for you?' I asked him.

'You have freedom of movement, I don't,' he said. 'You can go visit these people and ask questions, I can't.'

I still couldn't quite see where all this was headed.

Marckus took a moment before adding: 'I think it's 1939 all over again. I think we're poised on the brink of something. The physics is mind-boggling – Christ, I don't pretend to understand half of it myself – but what we're talking is huge.'

'Spell it out for me.'

'Bye-bye nuclear-power. Bye-bye rocket motors. Bye-bye jet-engines. If we can manipulate gravity, nothing will ever be the same. But expecting British Aerospace to develop this stuff is like asking someone who'd spent the first half of the 17th century building horse-drawn carts to come up with the outline for Stephenson's bloody Rocket. People don't really have a taste for it here, not at a senior level. They don't have the vision that the Americans do. Anti-gravity is too much mumbo-jumbo, not enough ROI.' He rubbed his thumb and forefinger under his nose and pulled a face. 'Return on investment – bean-counter talk.'

'So, if you can prove that the Americans really are up to something,' I said, 'you raise the stakes and get something going here? You get a real programme in this country with real money behind it?'

'Right – to a point,' he said. 'The difference is you're going to prove it.'

I stared at him, thinking I hadn't heard him right.

'It's not such a bad arrangement,' he said, reaching into the manila folder and pulling a wodge of papers from it. 'You get a short course in physics, a few tips on who to talk to, and maybe a story out of it.' He paused. 'The story of the biggest breakthrough in transportation technology since the invention of the wheel.'

'You said there were signs . . .' I said cautiously.

Marckus pushed the papers across the table. The cover-sheet was a photocopy from some kind of index. The writing was small, but my eyes were drawn to a section of text that had been highlighted with a yellow marker-pen.

I peered closer and words like 'magnetic monopoles', 'gravitational charge' and 'linear equations' swam into my vision.

'This is taken from something called the "Source Index",' he said. 'It's a list of doctorate papers published by universities around the world. The highlighted part shows you what is happening in one US university, that's all. There are dozens, maybe hundreds, of other facilities like these across America engaged in related research. All of them probing the sorts of areas you'd want to probe if you were pulling a field-dependent propulsion programme together. Some of the entries go back decades.'

Field-Dependent Propulsion; yet another term for anti-gravity.

'This thing is a multi-faceted, multi-layered puzzle and you won't solve it by picking away at the edges,' Marckus said. 'Begin with what you can see, then search for what you can't. What NASA's doing is in the open, so get over there and start looking and listening. Then go deeper. Some people say that the B-2 Stealth Bomber uses an anti-gravity drive system. It's like the bomb. The physics has been around for a long time. All it requires is some imagination, a shit-load of money and some technical know-how and someone who wants to put the pieces together.'

On the way back home, as the shadows lengthened and the traffic bunched on the choke-points into London, I thought long and hard about the road ahead and where it was taking me. I knew absolutely nothing more about Marckus than I had gleaned on him before the meeting. The meeting had confirmed that he was a highly gifted academic with strong ties to the British government and its defence industry and that he was unquestionably up to his eyes in secret work. But that was all.

It didn't make me feel particularly comfortable. But then again, I'd already painted myself into a corner and this, at least, was a way forward. Doing nothing was not an option if I wanted to progress things.

Marckus had given me the names of two people I needed to see in the States. One was Garry Lyles at NASA; the other was Hal Puthoff.

In American defence procurement, there was a white world and a black world – one that was open to public scrutiny and one that was so firmly closed, from an outsider's perspective it was like it didn't exist.

The white world had know-how, a good sprinkling of imagination and not a little money at its disposal.

The black world, on the other hand, had all of the above in spades.

Lyles clearly worked in the white world. But with Puthoff, it was absolutely anyone's guess.

Chapter 9

A larger than life depiction of Wernher von Braun, whose self-assured good looks conjured up images of post-war Hollywood matinee idols, stared down at me from a mural in the gallery above the arrivals hall at Huntsville International Airport.

Von Braun is still revered in this remote corner of Alabama as the founding father of the nearby George C. Marshall Space Flight Center, the NASA 'centre of excellence' for space propulsion. Von Braun had died more than 20 years earlier, but American rocketry remained as a testament to the work of Operation Paperclip, the USA's clandestine effort after World War Two to recruit Nazi V-2 specialists.

A complex of drab, rectangular buildings, 'the Marshall Center Wernher von Braun administrative complex', erected at that time, stands as testimony to the decade in which the high ambient humidity routinely reverberated to test-runs of engines for the Mercury-Redstone and Saturn V launch vehicles. The former took the first US astronaut, Alan B. Shepard, briefly into space three weeks after Yuri Gagarin in May 1961; the latter formed the basis of the giant Apollo moon-rocket.

Today, NASA Marshall scientists still work on the same basic technology that von Braun developed under the auspices of the Third Reich, even though chemical rockets, his great invention, are today regarded as too slow, too unreliable, too expensive and too inefficient.

It was mid-summer, four months after my meeting with Marckus, before I was able to wangle a trip to Marshall and do as he'd suggested.

Set in a vast area of grasslands, bordered by well-tended, lush forestation, Marshall was melting under the high-heat when I eased my rental car up the long approach road that peeled off the interstate to the hub of the Center itself. As I got out of the car, a military Huey thudded low overhead, disappearing behind the trees in the direction of the Redstone Arsenal, the Army's neighbouring missile test facility. As the low-frequency throb of the blades faded from earshot, it was replaced by the gentle swish of grass-sprinklers and the chatter of cicadas.

I was met at the entrance of the main building of the complex by a woman from public affairs. As we moved through the lobby, she sketched in some of the background to the programme I'd come to talk about with its manager, Garry Lyles.

Under the Advanced Space Transportation Program, ASTP, officials at NASA Marshall have been working since the mid-1990s on methods to reduce the cost of space access. Using today's generation of launch vehicles – the Space Shuttle Orbiter, the Delta and Atlas rockets – it costs around $10,000 per pound to put a satellite into low earth orbit. One short-term payoff of ASTP will be the development of a family of reusable rocket-planes that should cut these costs by a factor of ten by the end of the present decade. A mid-term goal, due to yield results in around 2015, is to develop a hybrid 'combined cycle' engine – half-rocket, half-jet engine – that will reduce launch costs by a factor of a hundred. Even though the hybrid engine, if it can be mastered technologically, will give NASA space access for as little as $100 per pound, it will never lead to anything other than a delivery system to low earth orbit.

For missions beyond Earth, scientists will either have to rely on improvements to the chemical rocket or on something radically different.

For a mission to the planets of the inner solar system a nuclear rocket could be developed in a relatively compressed timescale to cut round-trip journey times from Earth to Mars from many months to possibly just weeks. The feasibility of nuclear rockets had been proven in two NASA-developed solid core fission reactor test programmes, NERVA and ROVER, between 1959 and 1972. These never came to fruition, my PR guide told me, due to concerns over safety and also because emerging and more advanced nuclear fusion technology was seen as yielding three times the efficiencies. Like the fission rockets, however, fusion systems never happened either – not in the white world, anyhow. In the early 1990s, the Pentagon was reported to be working on a classified nuclear thermal rocket under the codename 'Timberwind', but soon afterwards it was cancelled. Since the black world is publicly unaccountable, it is impossible to know whether this really was the case.

Lightweight nuclear fusion propulsion is one methodology being studied for a manned NASA mission to Mars, my guide continued as we drifted towards the elevators, alongside 'light-sails', 'magnetic sails', 'anti-matter fusion drives', lasers and other exotic-sounding technologies.

While the physics behind these ideas is generally well understood, she added, the challenge in turning them into hardware is enormous.

But ASTP did not end there. In addition to examining ways to provide cheaper access to low earth orbit and coming up with propulsion systems that by the second half of the century will lead to 'the commercialisation of the inner solar system', its original remit went a lot further.

Into an elevator, half a dozen floors up, and we emerged into a world of bright walls, carpeted floors, posters of exotic spacecraft and artist's impressions of planetary systems, constellations and nebulae.

Up ahead, through an open door, sitting behind a large desk, Garry Lyles – bespectacled, bearded and thoughtful-looking – was waiting.

'Within a hundred years,' he told me over coffee, his soft southern drawl catching with emotion, 'we'll be going to the next star system.'

He spoke about the journey with such absolute conviction that it was difficult, in spite of the sheer fantasy at the heart of the idea, not to believe him. Lyles was a dreamer, but his dreams were rooted in work that was going on within the Wernher von Braun administrative complex as we spoke. ASTP scientists at Huntsville – people working right on the edge of what was known and understood in the harsh world of propulsion physics – were collaborating with other NASA facilities on science that would enable astronauts to make a 'fast trip' to the Alpha Centauri star system, Earth's nearest stellar neighbour, 4.3 million light years away. As head of this portion of ASTP, Lyles was bound up in the immensity of the challenge.

I asked what he meant by a 'fast trip'.

'We have set a goal to go to Alpha Centauri in 50 years – a 50-year trip.' He stared at me intently as he rocked back in his chair. 'You have to really use your imagination to come up with a propulsion system that will do that.'

'When will you be able to make that trip?' I asked.

He gestured to another part of the complex visible beyond the blinds of his window. 'We have a group of thinkers out there that's beyond anti-matter. They're looking at space-warp, ways to get to other stars in a Star Trek-like fashion. If we're not thinking about that kind of thing today, we won't be going to the stars in a hundred years' time. But I have to tell you that if you talk to these people, these engineers who are working on interstellar missions, they think they can do it. And I have no doubt that they're right.'

The trip, he reiterated, would be possible before the end of the 21st century.

The phone rang and Lyles took it, allowing me a moment to consolidate my thoughts. His team, he said, had opted for a 50-year trip because it represented a time-span most engineers could get their heads around. Fifty years, however, made no sense from the point of view of a practical mission, because even if it could be accomplished technologically, old age (or boredom) would have killed most of the crew long before they ever got there, let alone back again.

But that wasn't the point. The 'extra–Solar System' portion of ASTP was about getting scientists to think about realising impossible journeys. In 1960, there were those who said that landing a man on the moon by the end of the decade would be impossible. That we weren't colonising Mars by the end of the 20th century, they maintained, was directly attributable to the drop-off in research activity that followed the Apollo programme. If we wanted to get to the stars by the end of the 21st, Lyles said, we needed to be thinking about it now.

Which, though he didn't know it, was why I was here.

In 1956, a group of engineers had talked openly about an impossible idea. Lyles' 'fast trip' to Alpha Centauri was so impossible, his team was having to really reach – to think way out of the box – to come up even with half-answers to the challenge. To make the fast trip happen, NASA had no choice but to tap heavily into some taboo areas of science. That was my lead.

Supposing for a moment that NASA could come up with a vehicle capable of light-speed, a one-way trip to Alpha Centauri would take 4.3 years. Since the speed of light was a constant 186,000 miles per second, by extrapolation, the spaceship would be moving at almost 670 million miles per hour.

The 50-year journey time reduced the requirements to a more manageable 16,000 miles per second or 57.6 million miles per hour.

I had only once come across real hardware (aside from a laser, the ultimate 'speed of light' weapon) about which comparable speeds had been mentioned.

Shiva Star, the 250-ton capacitor I'd seen at Kirtland Air Force Base in New Mexico in 1992, had had a remit to fire a plasma bullet at 10,000 km per second.

And then, just a few years later, when I returned to Kirtland, it was like the programme never existed.

In the carefully regulated temperature of Lyles' office, I shivered.

Shiva's plasma bullet, designed to be dumped into the soft electronics of an incoming nuclear warhead at one thirtieth the speed of light, was but one of a handful of deep black programmes I'd encountered in the course of my career.

How many more secrets existed below the waterline of accountability? What other secrets were out there? What was the ultimate weapon? The questions churned, as they always did in a place like this, almost too fast for me to get a hold on them.

What Lyles and his counterparts in the black world shared was their absolute belief that they could engineer things other people said were impossible. It was a part of the feeling I got when I visited facilities like this and it thrilled and scared me with equal measure.

Why had Marckus sent me here? What did Lyles know?

When he got off the phone, I asked Lyles if he ever felt NASA was developing technology that had previously been developed by the Air Force in the black. He took a moment before answering, choosing his words carefully. I thought that by tweaking his professional pride I might be able to ruffle some of that soft, southern *savoir-faire*. But Lyles had seen me coming. 'Yeah, sometimes we do. But you have to understand that going into space is a very hard thing to accomplish. It takes a lot more than money. I would guess – certainly when it comes to single-stage-to-orbit technology – that NASA is at the cutting edge.'

This was the essence of the multi-layered conundrum that Marckus had touched upon during our meeting on the estuary. While NASA was struggling to develop technology that would reduce the cost of space access, there were those that said that the Air Force had already solved the problem. Believers in Aurora, the large triangular aircraft supposedly sighted over the oil rig in the North Sea, maintained that at least one variant was capable of conducting single-stage missions to low earth orbit. If, as George S. Trimble and his contemporaries had stated back in 1956, the US was on the threshold of developing anti-gravity technology, why spend billions on something as conventional (by comparison) as Aurora 20 years later? Or any other advanced kind of weapon system for that matter?

But when Lyles said that he doubted the Air Force had cracked single-stage-to-orbit, I actually had no way of knowing what he really knew. There was every chance that NASA was simply repeating work the Air Force had already carried out – and this applied to all aspects of its work 'on the cutting edge'.

I knew of several Congressmen who were deeply pissed off about it. On their watch, they'd committed money to a number of NASA programmes knowing damn well, but unable to prove it, the Air Force had duplicate technology up its sleeve.

A number of NASA scientists I had spoken to in other fields of endeavour had acknowledged a feeling – though it was often no more than this – that their work was a mirror-image of some hidden effort, something they couldn't see, that was going on in the black. If NASA scientists, therefore, were thinking 'way out of the box' to come up with science that would enable them to reach another star-system by the end of the century, maybe someone, somewhere had experienced that same feeling.

If anyone at Marshall had sensed such a presence, it would have been George Schmidt. Lyles recommended I go talk to Schmidt, because Schmidt, being a project engineer, was right on the cutting edge. A big, genial man with a football player's physique and a marine-style haircut, Schmidt, as head of propulsion, research and technology on the ASTP effort, waded routinely through the theory that could one day transform Lyles' interstellar vision into real hardware. Still under escort, I found him down a maze of corridors in a different part of the von Braun administrative complex.

Schmidt punctuated his remarks with references to 'superluminal velocities' – faster-than-light speeds – and 'space-time distortion', phenomena that weren't even accepted by mainstream physics ('although General Relativity and Quantum Mechanics don't discount the possibility'). By the same token, he said, there were some holes in current thinking – 'some windows of opportunity we may be able to take advantage of' – that might transition the theory sooner rather than later. One of these related to so-called 'worm-holes', short-cuts through the universe that could be created with large amounts of 'negative-energy density matter': material that existed only on the very fringes of theory.

Since it was known from the established physics of General Relativity that gravity, electromagnetism and space-time were inter-related phenomena, he said, it followed that any distortion of space-time might well yield an anti-gravity effect.

Questioned on the principles of a worm-hole, Schmidt quickly warmed to his theme. 'It's based on the idea that you can actually curve space,' he told me. 'Einstein discovered that large masses curve space. Well, it's the same thing with a worm-hole. You're curving and distorting space so that you don't have to follow the usual dimensions – you can go up and out, so to speak, into another dimension, if you will. And that allows you to travel much faster than the speed of light.'

I asked him if he ever got the feeling that anyone was working on this stuff elsewhere; that there may even have been breakthroughs – ten, twenty, maybe forty years earlier. His reply was more candid than I had anticipated.

'That's always a possibility. I'm not one for conspiracy theories, but we've learned about things that governments and organisations have been involved in in the past.'

He paused and rubbed the bristle hairs at the back of his head. I waited.

'It would be extremely frustrating to think that we're duplicating what was done maybe decades before. We like to think we're on the cutting edge, at the forefront of knowledge. It would be very discouraging to think that everything had been determined before and we were just sort of . . . here for show. I hope that's not the case. But you never know. Maybe in some aspects it is.'

It was then that my PR escort interjected.

'You know,' she said, 'I really don't think this is anything that we can comment on in any depth.'

I knew that the topic of conversation had taken an unwelcome turn. Now, I had intruded far enough.

'I'm simply trying to get a feel for the kind of capabilities that are out there – on both sides of the fence,' I replied.

'Then maybe you should go talk to the Air Force,' she said crisply. In any case, I was talking to the wrong people at NASA, she added. Huntsville was where the big picture behind the interstellar vision came together. But, she said, the 'breakthrough physics' that would one day lead to the engineering of a 'device' was being managed by the NASA Glenn Research Center at Lewis Field in Cleveland. It was from there, not here, that NASA was actively in the process of handing out contracts that aimed to prove or disprove whether there was any substance to the theory.

If I wanted to know whether the physics was doable, get up there, she told me. Talk to a guy called Millis. Marc Millis.

Before I left, I had one last question for Schmidt. I asked him if he genuinely believed that a worm-hole might one day give man the ability to cross the universe.

'Yeah,' he said, smiling but serious at the same time. 'If you can build a worm-hole, certainly. Who knows? But you've gotta be careful to make sure you've got the right end-point. Who knows where you'll end up if you just generate it? You could end up anywhere in the universe. So that's something to be very careful about.'

Six hundred miles to the north-east in Cleveland, Ohio, but a world away culturally, Marc Millis had indeed just received a half-million-dollar contract to test whether there was any theory out there, anything at all, that might one day enable a propulsion device to be built with the energy to take man out of the Solar System and into interstellar space.

Millis ran the NASA Breakthrough Propulsion Physics programme, referred to in the trade as BPP.

'This is it,' he told me, gesturing to the modest office he shared with another NASA official whose head was just visible on the other side of the screen that set the boundaries of Millis' work space. It had taken some hard negotiating after Huntsville, but I had managed to secure a visit to the Glenn Research Center, NASA's 'center of excellence for aero-propulsion research and technology', on my way to Washington, where I was due to conduct some interviews for the magazine before heading on down to Puthoff's place in Austin, Texas.

With his half a million dollars, Millis had awarded five contracts, any one of which might lead to the breakthrough that gave BPP its name.

The first contract, placed with Washington University, Seattle, was designed to test theory which maintained that a change in inertia was possible with a sudden energy-density change, induced, say, by a massive jolt of electricity. Manipulate inertia – an object's innate resistance to acceleration – and apply it to a spacecraft and you had the ability to reduce, if not negate altogether, its need for propellant. If the sums were right, it would continue to accelerate until it reached light-speed.

The second experiment was tasked with examining the reality of the so-called zero-point energy field and whether it existed at anything like the magnitudes that had been postulated by some scientists. If these people were even half right, Millis said, zero-point energy held enormous significance for space travel and offered entirely new sources of non-polluting energy on earth.

What, I asked, was zero-point energy?

For some years, Millis replied, there had been a developing under-standing that space was not the empty vacuum of traditional theory, but a seething mass of energy, with particles flashing in and out of existence about their 'zero-point' baselines. Tests indicated that even in the depths of a vacuum chilled to absolute zero (minus 273.15°C) – the zero point of existence – this energy would not go away. The trouble was, no one knew quite where it was coming from. It was just there, a background radiation source that no one could adequately explain. With millions, perhaps billions, of fluctuations occurring in any given second, it was theoretically possible to draw some – perhaps a lot – of that energy from our everyday surroundings and get it to do useful work. If it could be 'mined' – both on Earth and in space – it offered an infinite and potentially limitless energy source.

The third experiment, he continued, would set out to examine the unproven link between electromagnetism and gravity and its possible impact on space-time. Perturbations of space-time not only promised a propulsive anti-gravitational effect, but opened up other more esoteric possibilities, such as time-travel.

I asked Millis if this stuff was for real and for a moment he looked unsure. The mere mention of time-travel and we were suddenly into science-fiction territory. It was enough to put any serious scientist on the defensive.

'The immediate utility is not obvious,' he said stiffly, 'but it's like a foot in the door.'

The fourth experiment looked at 'superluminal quantum tunnelling' – faster-than-light speed. Recent laboratory tests had shown light pulses accelerating beyond the speed of light, thereby shattering Einstein's Theory of Relativity, which said that the light-speed barrier was unbreachable. If a light pulse could arrive at its destination before it left its point of departure, the theory said a spacecraft might be able to as well.

'How it translates into space travel,' Millis said, anticipating my next question, 'is way down the pike.'

Again I was presented with a failure of fit. If NASA was still struggling with the theory now – 40 years and a world of knowledge after the pronouncements of George S. Trimble and his colleagues in 1956 – what on earth had led them to believe they could crack gravity then?

Something I'd written down tugged me back to the present. I flicked back over the list of experiments and did a quick inventory count.

'You said there were five contracts. I've only got four. What's the fifth?'

He looked momentarily taken aback.

'It's conducted under the auspices of the Marshall Space Flight Center itself,' he said. 'Since you'd just come from there I rather presumed they'd briefed you on it.'

I shook my head. 'Briefed me on what?'

Millis reached into a drawer and pulled out a file full of newspaper cuttings and magazine articles. He rifled through them until he found what he was looking for, a cutting from Britain's *Sunday Telegraph* dated the 1st September, 1996. The headline read: 'Breakthrough As Scientists Beat Gravity'.

'I'm surprised you didn't know about this,' he said, pointing to its source.

I said nothing. I had a photocopy of the same story in my files at home. I'd not taken much notice of it because it had seemed totally outlandish.

The story detailed the experiments of a Russian materials scientist called Dr Evgeny Podkletnov, who claimed to have discovered an anti-gravity effect whilst working with a team of researchers at Tampere University of Technology in Finland. What made the claim different from 'so-called "anti-gravity" devices put forward by both amateur and

Breakthrough as scientists beat gravity

by ROBERT MATTHEWS
and IAN SAMPLE

SCIENTISTS in Finland are about to reveal details of the world's first anti-gravity device. Measuring about 12in across, the device is said to reduce significantly the weight of anything suspended over it.

The claim — which has been rigorously examined by scientists, and is due to appear in a physics journal next month — could spark a technological revolution. By combatting gravity, the most ubiquitous force in the universe, everything from transport to power generation could be transformed.

The Sunday Telegraph has learned that Nasa, the American space agency, is taking the claims seriously, and is funding research into how the anti-gravity effect could be turned into a means of flight.

The researchers at the Tampere University of Technology in Finland, who discovered the effect, say it

HOW THE ANTI-GRAVITY DEVICE WORKS

A 1-stone object would lose 4.5oz of its weight through 'shielding' effect of the device

Whole assembly cooled with liquid nitrogen

Ring of super conducting ceramic (Yttrium-barium-copper oxide), spinning at 5,000 rpm

Solenoids used to put magnetic field around ring

Three solenoids, which allow ring to levitate magnetically

Unit approx. 12in diameter

tures. The team was carrying out tests on a rapidly spinning disc of superconducting ceramic suspended in the magnetic field of three electric coils, all enclosed in a low-temperature vessel called a cryostat.

"One of my friends came in

the *Journal of Physics-D: Applied Physics*, published by Britain's Institute of Physics.

Even so, most scientists will not feel comfortable with the idea of anti-gravity until other teams repeat the experiments.

professional scientists,' wrote the paper's science correspondent, was that it had survived 'intense scrutiny by sceptical, independent experts'. It had been put forward for publication in *Journal of Physics D: Applied Physics*, published by Britain's Institute of Physics.

When the *Telegraph* hit the streets, Podkletnov's world turned upside down. He was abandoned by his friends, ostracised by the university, which claimed that the project did not have its official sanction (and later threw him out), and the journal pulled his paper. Leaving his wife and family in Tampere, Podkletnov returned to Moscow to lick his wounds until the furore had died down.

And all because the *Sunday Telegraph* had used a single, taboo word in its opening sentence: anti-gravity. The term may have been vital for 1950s pulp sci-fi, but it had no place in the world of mainstream physics.

What was unfair, was that Podkletnov had been so careful to avoid the term himself. All along, he had referred to the test as a gravity *shielding* experiment.

It was still heresy; and he had burned for it.

But here was Millis saying that NASA *had* taken Podkletnov's claims seriously. Not only that, but NASA had been apprised of the 'gravity-shielding' properties of superconductors even before the *Sunday Telegraph* story broke.

I wondered why no one had told me about all this while I was in Huntsville.

In 1993, the Advanced Concepts office at the Marshall Space Flight Center was handed a copy of a paper written by two physicists, Douglas Torr and Ning Li, at the University of Alabama at Huntsville. It was called 'Gravitoelectric-electric coupling via superconductivity' and predicted how superconductors – materials that lose their electrical resistance at low temperatures – had the potential to alter gravity. Even if the amounts were fractional, NASA officials quickly saw the benefits. Applied to something like a rocket, even fractional methods for cutting launch-weights would, over time, lead to a considerable reduction in the space agency's fuel bills.

In their paper, Torr and Li went out of their way to avoid the term 'anti-gravity'. One whiff of it, they knew, and no one would take their work seriously.

It worked. NASA promised money for an experiment.

But when the Podkletnov story broke, Millis told me, NASA decided to take a multi-path approach instead. First it tried to contact Podkletnov to see if he would share his findings with them. When the Russian refused, it elected to do the experiments itself, setting up the 'DeltaG group' at Marshall specifically to look for interactions between super-conductors and gravity.

At first, Ning Li, the brilliant Chinese woman scientist who had predicted a one per cent weight change in her original calculations, joined forces with the DeltaG effort. Then, apparently frustrated by the pace of work at Marshall, she set up to pursue the experiments on her own. If she was getting anywhere with them, she wasn't saying, all her most recent tests having been conducted behind closed doors.

Since that time, according to Millis, the DeltaG group had found cause to believe that *something* unusual was happening when gravity fields were influenced by superconductors, although the precise nature of the effect was a matter of conjecture. Podkletnov had claimed that two per cent weight reductions were being consistently observed during his experiments – a result, if you could call it that, that was so close to the Torr/Li prediction it was scary.

Like the whirling gyroscopes investigated by BAe, the low percentages were, in a sense, immaterial. What was being talked about here – and apparently engineered by Podkletnov – was a phenomenon that science said was impossible.

The DeltaG experiments at Marshall continued, but, without Podkletnov's input, Millis admitted that things had not gone as fast as NASA would have liked. Part of the problem had been its inability to reproduce superconductors of the size and fidelity Podkletnov said was required to achieve gravity shielding – a term with which even Millis, who was broader-minded than most, was uncomfortable.

'It's more accurate to call it an anomalous gravitational effect, but even that is jumping the gun.' It could, he said, simply be an experimental error.

But what if it wasn't? If Podkletnov was right, if he had found a way of altering an object's weight, aerospace was about to change – civilisation, too, maybe – and forever. Conscious, no doubt, that it might be staring at the greatest scientific discovery since the bomb, I could see why NASA was hedging its bets.

It had solicited a company in Columbus, Ohio, to try to develop superconducting ceramic discs like those built by Podkletnov. Without the Russian's secret recipe, however, it hadn't been getting on as well as it might. At the time of my visit to Ohio, the Columbus-based company had managed to build small discs, but it still hadn't acquired the knack for turning out the larger variety – the type that Podkletnov said was crucial to the success of the experiment. And this was the bitter-sweet irony. Being a materials scientist, Podkletnov had an insight into the way the discs were constructed that left NASA, with its $13 billion annual budget, floundering in his wake. Tempting as it was to view Huntsville's silence over Podkletnov as a deliberately misleading omission, I suspected that NASA's impotence in the face of this irony was the real reason no one had mentioned him.

On the flight down to Texas, I picked through a wodge of cuttings I'd managed to track down during my brief stop-over in Washington. From the little that was known of Podkletnov's Tampere experiments, it was apparent that the superconductors needed to be spun at around 5,000 rpm inside a large steel container filled with liquefied gas for cooling. Electromagnets were used to make the superconductors levitate and get them up to speed. In the absence of any firm data on Podkletnov's discs, though, attempts by NASA to duplicate his work were inconclusive as far as gravity shielding was concerned. A couple of times, the readings had blipped – a little like Dr Evans' experience with the Kidd inertial thrust machine at BAe – but they were never consistent enough to be taken seriously.

If Podkletnov knew all this, I thought – and somehow or other, I guessed he probably did – then he was more than likely pissing himself laughing.

Either he had perpetrated a fantastic confidence-trick, or he really had pulled off one of the greatest scientific discoveries of the 20th century.

So, where the hell was Podkletnov?

The best and only clues to his whereabouts had been provided by an American journalist called Charles Platt who had documented his attempts to track down Podkletnov in a lengthy article for *Wired*, a highly respected magazine for computer-minded tech-heads. After the *Sunday Telegraph* story, Platt managed to get a phone number for Podkletnov and they spoke a couple of times, but only on the basis that Platt wouldn't publish anything without the Russian's consent.

'He told me how he made his discovery,' Platt wrote more than a year later, quoting the Russian: '"Someone in the laboratory was smoking a pipe and the pipe smoke rose in a column above the superconducting disc. So we placed a ball-shaped magnet above the disc, attached to a balance. The balance behaved strangely. We substituted a non-magnetic material, silicon, and still the balance was very strange. We found that any object above the disc lost some of its weight, and we found that if we rotated the disc, the effect was increased.'"

When Platt suggested that huge amounts of energy would be required for this to be remotely so, Podkletnov reportedly became irritated.

'We do not need a lot of energy,' he'd snapped back. 'We don't absorb the energy of the gravitational field. We may be controlling it as a transistor controls the flow of electricity. No law of physics is broken. I am not one crazy guy in a lab; we had a team of six or seven, all good scientists.'

And that, for the next year or so, was that. It was only when an intermediary told Platt that Podkletnov had returned to Finland, where he was now set up as a materials scientist in a local company, that the

American jumped in a plane and finally managed to confront the man who'd dared to claim he had shielded gravity.

What Podkletnov had been doing during his year-long self-imposed exile in Moscow was never made clear, although the Russian soon hinted that he had not been idle. He was quick to explain to Platt that his original findings at Tampere had been meticulously charted using a mercury barometer. Immediately above the superconducting disc, it had registered a 4 mm drop in air pressure, because, the Russian said, emphasising the point, the air itself had been reduced in weight.

When he'd taken the barometer upstairs, Podkletnov had found the same drop in pressure over the point where the experiment was taking place on the floor below. He went up to the top floor of the building and it was the same thing.

This showed that gravity reduction would not diminish with distance, that the effect had no limit. Podkletnov's gravity shield went on, extending upward in a 30 cm diameter column – forever.

And then, if that wasn't good enough, pay-dirt. The two per cent weight reduction in all the air above the disc meant that a vehicle equipped with gravity shielding would be able to levitate, buoyed up by the heavier air below.

'I'm practically sure,' Podkletnov had told Platt, giving him an intense look, 'that within ten years this will be done. If not by NASA, then by Russia.'

During his year in Russia, he went on to reveal, he had conducted research at an unnamed 'chemical scientific research centre' where he had built a device that *reflected* gravity. By using superconductors, resonating fields and special coatings – and 'under the right conditions' – gravitational waves had been repelled instead of blocked. Applied to an air vehicle, Podkletnov said, this 'second generation of flying machines will reflect gravity waves and be small, light and fast, like UFOs. I have achieved impulse reflection; now the task is to make it work continuously.'

Their meeting over, Podkletnov slipped back into the shadows and Platt returned home. While there had been sightings of Podkletnov in the years since – reports of his whereabouts had surfaced in Japan, Russia and Finland – he had, to all intents and purposes, disappeared again.

Back in the States, Platt canvassed the physics mainstream for its reaction to the Podkletnov claims. Most scientists laughed outright. Gravity shielding, they said, was simply inconceivable.

I made a note on the flight home to ask Marckus to keep his ear to the ground for signs of the Russian's whereabouts. It was tempting to believe that Podkletnov was by now working in the bowels of a crumbling military research institute somewhere deep behind the Urals, refining his experiments to the accompaniment of a howling Siberian gale. And maybe this wasn't a million miles from the truth.

But scientists had a habit of talking to one another, whatever obstacles seemed to separate them. And since Podkletnov had shown himself to be loquacious enough when it suited him, I thought he'd probably surface again before too long.

When he did, I needed to be there.

I was intrigued by the Russian's work. It echoed something that Professor Young had said in his lecture on anti-gravity – something that had sounded right at the time. If gravity control is ever to be discovered, Young had said, it would most likely be as an unexpected by-product of work in some completely different field. An accident.

Podkletnov hadn't been looking for an anti-gravity effect. He'd been playing around with superconductors. If he'd been working in a Western laboratory, his discovery might have forever gone unnoticed. But in Russia – and Finland, too, apparently – it was OK to light up a pipe in a room full of test equipment. And, bingo, the smoke had hit the column of shielded air and snaked straight up to the ceiling.

It struck me as not impossible that Podkletnov had discovered – in a million-to-one fluke – what Trimble et al had predicted theoretically in 1956.

Had Podkletnov tripped over the answer and got in by the back door?

If anyone knew it was Dr Hal Puthoff, the second of the two contacts passed to me by Marckus during our meeting by the estuary.

Even in the cocooning environment of the aircraft, nursing a stiff drink, I felt anxious at the prospect of meeting him.

Up until now, I'd managed to palm off my interest in NASA's advanced propulsion activities as a quirky sideline to my more regular journalistic pursuits. I knew, though, that when I pitched up at Puthoff's door, a discreet address remote from any other legitimate aerospace business, there'd be no hiding what I was really up to.

Chapter 10

I reached Puthoff's offices after a drive through the industrial suburbs of the Texas state capital, parked up and double-checked that I had the right address. I had expected the Institute of Advanced Studies to be housed in another windowless science block, a modern version of the concrete edifices at NASA, replete with clean rooms and scientists swishing around in white coats. But the IAS was a small glass-fronted office in a leafy business park. Several of the other lots were vacant and at least one was boarded up.

Puthoff's reputation as a scientist of vast repute went a long way before him. His papers on gravity and inertia were viewed as seminal works by the science community. He also sat on something called the Advanced Deep Space Transport group, ADST, a low-profile organisation composed of people that someone had described as being 'on the frontier of the frontier'. Staffed by select NASA, Air Force and industry representatives, it was the real cutting-edge, but you had to know where to look for it.

Puthoff's link to NASA was bound up in theoretical work that he had been conducting since the early 1970s into the zero-point energy field.

Before getting out of the car, I was beset with a rush of all the old doubts. Did I really want to declare my interest in anti-gravity to a man who had clear connections to the intelligence community?

I checked my watch. It was a little after five-thirty in the morning back in England, but I needed reassurance. I called Marckus on my cellphone.

He picked up on the second ring, sounding bright and alert. I wondered for a moment whether Marckus ever slept, then dispensed with the small-talk and told him of my concerns.

Marckus, showing a tendency that would become familiar over the course of our acquaintance – a tendency to impart knowledge well after the time I needed it most – informed me that NASA, being a government institution, was, in effect, the outer marker of an elaborate but invisible security system.

'How's that?' I asked.

'You show up asking the kind of questions you asked at Huntsville and Cleveland and, believe me, someone, somewhere will have registered the fact. I have little doubt that you began pulsing away on radar-scopes at government agencies too numerous to mention the moment you even logged your requests to visit NASA. You've already crossed the threshold. It's too late to go back now.'

I felt the muscles in my stomach tighten, but I said nothing. This propensity to unsettle was also a part of Marckus' act. I asked him instead, as casually as I could, for a data-dump on everything that was known about zero-point energy.

I'd already grasped the principles: that space – the air that we breathed, the atoms in our bodies, the far reaches of the cosmos, everything – was filled with a churning 'foam' of energy. Its presence was signified by the background static hiss of the universe – something you could hear on a transistor radio – but as far as visual detection went, was invisible. Which was what made it so controversial.

In 1948, Marckus told me, a Dutch physicist, Hendrik Casimir, came up with a theoretical model for proving that the 'quantum vacuum' existed. If two aluminium plates were placed very close together, so close that their separation was less than the wavelengths of the fleeting particles in the quantum foam, it followed, Casimir reasoned, that there would be nothing between the plates.

Because the theory said that everything outside the plates would still be seething with zero-point fluctuations, the external force pushing in on the plates ought to be enough to close them together, thereby proving the existence of the ZPE field.

The difficulty, however, related to the experiment. To exclude the wavelengths of the fluctuating particles, Casimir's plates had to be exactly parallel and separated by less than one micron – one hundredth the width of a human hair. It wasn't until 1997 that science was able to come up with equipment that could satisfy the tolerances required for the Casimir experiment. When it did, it found that the Casimir Effect existed exactly as predicted.

And with it flowed the corollary that the quantum vacuum, zero-point energy – call it what you will – existed as well.

In principle, there was no reason why mechanisms couldn't be built to tap into this energy.

And since the fleeting particles were popping in and out of existence billions of times every second, on every conceivable frequency, in every possible direction, there was, theoretically speaking, no limit to the amount of energy resulting from the fluctuations.

In principle. Theoretically speaking.

'The argument now is not about whether the ZPE field exists,' Marckus concluded, 'but about how much energy it contains. It comes down to a difference in people's calculations and interpretations. Some say that there's enough energy in a shoebox to blow the world apart; others say that all the ZPE in the volume of the earth couldn't boil you an egg with it.' He paused. 'I think you'll find with Puthoff that it's closer to the shoebox scenario.' And then he signed off.

I locked the car, walked over to the IAS office, pushed the glass and watched as it swung open. There was no one in the front office, so I called out a couple of times and, not getting any answer, moved towards the back of the building. I followed a corridor till I heard a man's voice. Poking my head around a door, I saw Puthoff sitting on the edge of his desk, a phone jammed to his ear. He was shorter than he'd appeared in photographs, but his puckish features and thick head of hair belied his 60-odd years. He waved me in, pointed with a mock grimace to the voice on the other end of the receiver, and gestured for me to grab a seat. A couple of minutes later, he hung up, muttered something about a cancelled meeting in Washington, and asked if I'd like a coffee.

While we were standing by the coffee-machine, I asked him which of Marckus' analogies he inclined towards.

'If we're right,' he said, as I stirred creamer into my coffee, 'and you have all possible directions of propagation in the universe, and all possible frequencies and wavelengths, and every one of them has this little contribution of zero-point energy, when you add it all up, the numbers are ridiculous.'

How ridiculous? I asked.

He paused and added matter-of-factly: 'There's enough energy in the volume of your coffee-cup to evaporate all the world's oceans many times over.'

He showed me a large workshop at the back of the building filled with workbenches, electrodes and measuring equipment. This, Puthoff said, was where he and his small team worked on the ZPE devices that people routinely sent him for analysis; tiny little reactors that supposedly drew energy out of the vacuum of space: the space that we occupied and the space, in its quantum sense, that occupied us.

The machines came in all shapes and sizes. Some were solid-state devices with no moving parts that used the pinching motion of the Casimir effect to create clusters of electrical charge with energy outputs many times greater than the kick they required to get going.

Others used the millions of imperceptible yet infinite movements of the quantum foam to stimulate a vibrating motion in magnetic fields. These vibrating fields, their proponents claimed, served as 'gates' via which unending supplies of vacuum energy could be fed through electronic circuits and put to use.

If they could be proven to work repeatedly and reliably, Puthoff said, imagine it: no more pulling power off the national grid. You'd have a reactor in a box in your backyard; a little thing the size of a microwave-oven that sucked energy from the space that it occupied and belted out clean, unending power wherever it was needed. Initial machines would feed power to your house, but later, as they developed in sophistication, just as computers had developed exponentially at the end of the 20th century, these devices would shrink in size and double in output; and they'd keep on shrinking and doubling, over and over, until they'd be small and powerful enough to put in cars or aircraft or submarines. Applications would be limited only by the imagination of the user.

'If,' I said. 'You said, "if".'

'Ah, well, that's the point,' he replied. 'Of the 30-odd devices that have come through this door, none has passed the magic test yet: a demonstrable measurement of more energy flowing out than is flowing in.'

But it would happen. Maybe tomorrow, maybe in 30 years. But for sure, one day someone was going to announce that they'd invented a machine that could extract energy from the immeasurable pulse of the universe. It'd rate a paragraph in the newsbriefs columns, but clip it or keep the paper, mark it well, Puthoff said, because on that day the world would change. Nothing would ever be the same again.

As he talked, Puthoff filled in the gaps in the résumé that I'd pulled off the Net. Soon after gaining his master's degree at the University of Florida, he went on active duty as a Naval Intelligence officer seconded to the National Security Agency. When he left the Navy, he converted to civilian status at the NSA, then went on sabbatical at Stanford University in California to get his Ph.D. He gained his Ph.D in 1967, resigned from the NSA and soon afterwards joined the Stanford Research Institute, a spin-off from the university, set up, amongst other things, to pursue heavily classified research for the US defence and intelligence communities. Puthoff's thing was lasers, but it was at SRI that he developed the notion of the remote viewing programme for the CIA and the DIA. The RV work formed the hub of the 'heavily classified research' that he undertook at SRI for the next 13 years.

I asked how he'd made the jump from lasers to remote viewing.

'It sounds weird, but it was a very straightforward thing,' he said. In the early 1970s, physicists were searching everywhere for evidence of tachyons – particles that theoretically had the ability to travel faster than light. Finding them was problematic, because, if the theory was right, they couldn't travel below light speed, making identification somewhat challenging.

Someone had given Puthoff a book called *Psychic Discoveries Behind the Iron Curtain*, which mentioned how Soviet scientists were looking for evidence of ESP in plant-life. The Russians had connected up a pair of plants to polygraphs, separated them over huge distances and shielded them from all electromagnetic frequencies. They found that when one plant was burned with a cigarette, or even if someone was thinking about it, the other plant responded on the polygraph.

Since one of the postulated characteristics of tachyons was that they could penetrate all kinds of shielding, it occurred to Puthoff that one of the few places that scientists hadn't looked for tachyons was in 'organic living systems'. He wrote up a proposal that aimed to verify if plants were 'tachyonically-connected' by linking separated algae cultures with lasers. He sent the proposal to an acquaintance involved in similar work who happened to show it to Ingo Swann, a noted psychic and medium, who in turn approached Puthoff, assuming that he was interested in ESP. Even though he wasn't, Puthoff was sufficiently intrigued to invite Swann to SRI, where he proposed to immerse him in a 'shielded quark magnetometer', a 'tank' in a vault that was totally cut off from all known emissions and frequencies.

As Swann approached the vault, which he hadn't been forewarned about, he appeared to perturb the operation of the magnetometer. The dials and instruments inside it all flickered into life. He then added insult to injury by remote viewing the interior of the apparatus, 'rendering by drawing a reasonable facsimile of its rather complex construction', according to Puthoff, even though details of it had never been published anywhere.

Puthoff wrote up his findings, circulated the report among his colleagues and got on with other work.

A couple of weeks later, the CIA showed up at his door and invited him to set up a classified research programme to see if it was technically feasible to use clairvoyance to spy on the Soviet Union. It turned out that the CIA had been tracking Soviet efforts to do the same thing for several years, but didn't believe what it was seeing. It was only when Puthoff wrote up his paper that it put two and two together and thought it'd better set up something of its own. And so the US controlled remote viewing programme was born, with Puthoff as its director.

For the next decade and a half, US 'psychic spies' roamed the Soviet Union, using nothing more than the power of their minds to reconnoitre some of the Russians' most secret R&D establishments. If you believed what you read in published accounts of these people's activities, they were able to roam not merely in three dimensions – up, down, left and right – but in the fourth dimension as well: time. They could go back in time to review targets and they could look at them in the future as well.

Puthoff showed me a picture that one of his team of remote viewers had drawn of an 'unidentified research centre at Semipalatinsk', which had been deeply involved in Soviet nuclear weapons work. He then showed me some declassified artwork of the same installation, presumably drawn from a US spy satellite photograph. The two were damn near identical.

It all seemed a long way from zero-point energy, gravity and inertia, in which Puthoff had been busying himself even before he left the remote-viewing field.

Or was it?

In a recent paper written by Puthoff, which I had already seen, he'd mentioned that the Russian physicist and Nobel prize-winner Andrei Sakharov had published a paper in 1967 suggesting that gravity and inertia might be linked to what was then still a highly theoretical proposition: vacuum fluctuations of the zero-point energy field. Now that the zero-point energy field had been proven to exist, 'there is experimental evidence that vacuum fluctuations can be altered by technological means,' Puthoff had written in the paper. 'This leads to the corollary that, in principle, gravitational and inertial masses can also be altered.'

Puthoff was too smart to use the term – for all the reasons that Podkletnov, Ning Li and the others hadn't – but he was saying that anti-gravity was indeed possible.

And, so were the Russians. All you had to do – somehow – was perturb the zero-point energy field around an object and, hey presto, it would take off.

I asked him why NASA and the Air Force should be so interested in ZPE when it was obvious that there weren't going to be any practical applications of it for years – maybe centuries.

Puthoff looked at me in a meaningful way. 'Unless we find a shortcut.'

'Is that a possibility?'

'Podkletnov could have come across a shortcut.'

The maverick Russian materials scientist and the idea of pipe-smoke hitting a gravity-shield had entered the conversation so abruptly that it caught me off-guard. Puthoff was saying that anything was possible. And this from a man who had spent most of his professional life plugged into the heart of the classified defence environment.

I probed further, but gently. Were there already forms of aerospace travel out there – in the black world, maybe – whose principles contravened, if not the laws of physics, then at least our understanding of aerodynamics?

He sucked the top of his pen, giving the question a lot of thought before responding. 'I've certainly talked to people who claim that something is going on,' he said; pausing to add: 'I would say the evidence is pretty solid.'

I felt myself rock back on my chair.

I eased back on to the more comfortable subject of the NASA BPP study. 'Which of the five methods outlined in Breakthrough Propulsion Physics would you say is the one most likely to have a payoff?' I asked him.

Because of Puthoff's promotion of ZPE, I thought it inevitable that this explanation would have figured in his answer. But it didn't. Without even thinking about it, he said his money was on the third experiment – the one about perturbing space-time; anti-gravity from time-travel.

Given Puthoff's strong connections to the military-intelligence community, it was tempting to dismiss all such talk as a deliberate blind, something to lead me away from a more probable area of breakthrough. But it didn't seem that way at all.

I liked Puthoff. He seemed on the level. Besides, one thing I had learned after years of formal briefings and interviews was how difficult scientists found it to lie when confronted with a direct question on their chosen subject. Spooks could lie at the drop of a hat. It was what they were trained to do. But scientists found it inherently unnatural, since it contravened all their instincts to match a question quid pro quo with data.

Puthoff's answer to the question had been swift and visceral. And there had been something else, too; a look on his face, a willingness to confront my gaze, that in itself conveyed information.

We'd been talking about breakthroughs – events that, at a stroke, had the power to short-circuit the evolutionary development of technology.

The important thing about breakthroughs was that they didn't happen over an extended period of time. They changed things overnight. And they could occur at any moment.

It was only later, when I was back on the freeway, crawling south in the bumper-to-bumper traffic, that I had the feeling he'd been trying to tell me that some kind of payoff from the third experiment had already produced tangible results.

Chapter 11

If anti-gravity had been discovered in the white world, then someone, somewhere had to be perfecting it – maybe even building real hardware – in the black. The trouble was, I had no idea where to begin looking for it. Contrary to the implicit promise of revelation contained in Marckus' hints by the estuary, neither Lyles nor Puthoff had taken me any further in so far as hard and fast leads were concerned.

On my return, I tried to raise Marckus. I left several messages on his answer-machine. In the first of them I foolishly spelled out the fact that I didn't know a whole lot more – beyond strict theory, at least – than I had before my trip to the States.

During subsequent calls, as the answer-machine clicked in, I just knew that Marckus was there – listening in, but refusing to pick up. I was beginning to get the measure of this relationship. Marckus gave a little and I had to give something back. He now knew that I had returned from the States with nothing, so he was punishing me for it. Well, I didn't need to play psych-games with some spook academic with a warped taste for drama. I had enough on my plate already – a major feature article on stealth technology to deliver. Anti-gravity, for the time being, was going to have to go on the back-burner and Marckus could stew with it.

As background, I reread an account of what is probably the most definitive history of stealth to date: *Skunk Works*, the story of Lockheed Martin's super-secretive special projects facility, as told by the guy who had run it for 16 years, Ben Rich.

In the summer of 1976, America's leading radar expert, Professor Lindsay Anderson of the Massachusetts Institute of Technology, arrived at what was then the Lockheed Skunk Works in Burbank, California, with a bag of ball-bearings in his brief-case. The ball-bearings ranged in size from a golf ball to an eighth of an inch in diameter. Rich, a grizzled veteran of Lockheed's secret projects department for more than 25 years, led Anderson to a mocked-up design of a funny-looking diamond-shaped aircraft, poised on a plinth in the middle of a blacked-out hangar.

Over the course of the day, Anderson repeatedly attached the ball-

115

bearings to the nose of the aircraft, starting with the largest and ending with the smallest, while Lockheed engineers fired radar signals at it. What they were looking to measure was the radar cross section, or RCS, of the prototype aircraft, codenamed Have Blue. The moment the radar picked up the ball-bearing instead of the aircraft behind it was the moment its RCS could be accurately plotted.

When the test was over, Anderson zipped the bearings back into his case, left the site as quietly as he'd arrived and returned to MIT to process the data.

A few days later, Rich received a telegram from the air force chief of staff. Have Blue had been classified 'Top secret – Special Access Required', a security grading reserved only for 'quantum leaps' – programmes that jumped right over the state of the art. The first of these super-secret 'black' programmes had been the atomic bomb. The telegram had just pitched 'stealth' into the same category of secrecy.

Have Blue's RCS had bettered every single one of the ball-bearings Anderson had attached to it – even the one that was sized down to an eighth of an inch. This gave the aircraft and its operational successor, the F-117A Stealth Fighter, a radar signature equal to that of a bee-sized insect. Stealth, which up to that point had been viewed by sceptics as nothing more than puff and magic, suddenly went 'deep black' – so secret that no one outside the programme would ever know it was there. Lockheed had discovered a way of making aircraft invisible to radar, but the stakes were far, far higher even than that.

Soon after the late 1977 first flight of Have Blue from the US Air Force's remote testing base on the dry desert lake bed at Groom Lake, Nevada, also known as Area 51, Dr Zbigniew Brzezinski, President Jimmy Carter's national security adviser, flew into the facility in an unmarked jet and grilled Rich about the true significance of his firm's breakthrough. Rich put it on the line for him.

'It changes the way that air wars will be fought from now on. And it cancels out all the tremendous investment the Russians have made in their defensive ground-to-air system. We can overfly them any time at will.'

'There is nothing in the Soviet system that can spot it in time to prevent a hit?' Brzezinski asked, jotting notes as he talked and listened.

'That is correct,' Rich replied simply.

After more than three decades of stand-off between the two super-powers, the realisation made Carter's national security chief catch his breath. If the Russians wanted to match stealth with counter-stealth, theoretically they could do it, but it would bankrupt them in the process.

Stealth had the power to end the Cold War, but it also had the power in the meantime – should the Russians realise what the Americans were up to – to trigger World War Three.

Have Blue was a strange-looking aircraft. It didn't look mean and hungry. It didn't look like the kind of plane that could start and end wars. It had the appearance of a bunch of bolt-on geometric shapes that had somehow been hammered into the vague outline of a jet. It turned out there was more than a good reason for that.

Have Blue hadn't been the creative inspiration of an aerodynamicist, but of a mathematician.

Begin with what you can see, then search for what you can't.

As I wrestled with the stealth feature, Marckus' words from our meeting battered around my head, forming as a dark thought with a black body and soft wings.

I had trawled the white world and found five possible pathways to anti-gravity. None of them was visible in the developmental activities of the US aerospace industry, but just because you couldn't see them, it didn't mean they weren't there.

Just as it had, or hadn't been with stealth a decade and a half earlier.

Ben Rich and I had sparred on a number of occasions on the stealth question – most recently at an air show where he'd turned up, desperately ill with cancer, to promote his book. I'd respected him utterly and liked him hugely, sensing in his presence the grit and wisdom of a generation of post-war aviation pioneers that wouldn't be around us for much longer.

It was Rich who'd once told me of a place – a virtual warehouse – where ideas that were too dangerous to transpose into hardware were locked away forever, like the Ark of the Covenant in the final scene of *Raiders of the Lost Ark*. It had almost happened to stealth.

By starting with stealth, something that I knew about, something I could *see*, maybe I could pick up traces of that energy, the friction that astronomers looked for when hunting down black holes.

In the year ahead, the calendar of aerospace and defence events showed that I had multiple excuses for visiting the States – to look for the self-generated heat of a buried anti-gravity effort's interaction with the real world.

As I sat behind my desk, calling contacts and sources in an effort to plug holes in the incomplete, still highly secret history of stealth, I could hear the faint echo of a signal. It was almost impossible to decipher, but the cold side of my training, the part I kept separate from the muffled wing-beats of the creature let loose upon my thinking by Marckus, said anti-gravity had something to do with the shared origins of the two most powerful weapons of the 20th century – stealth and the atomic bomb.

Have Blue would never have happened, but for the work of a retired Lockheed maths-guru named Bill Schroeder and a Skunk Works software engineer called Denys Overholser. Schroeder revisited a set of mathematical formulae originally derived by Maxwell and refined by a turn-of-the-century German electromagnetics expert called Arnold Johannes Sommerfeld. Between them, these two physicists predicted the manner in which a given geometric shape would scatter or reflect electromagnetic radiation. A Russian, Pyotr Ufimtsev, honed their work into a more simplified set of equations in the early 1960s, but Ufimtsev could only apply the equations to the most basic geometric shapes. Schroeder's and Overholser's breakthrough had been to take Ufimtsev's concept and apply it to the inherently complex form of an aircraft, rejecting the regular smooth, curved lines of aerodynamic aesthetics in favour of a set of ugly flat angled panels. If an aircraft could be broken down into hundreds, maybe thousands of these two-dimensional shapes, if each flat surface or 'facet' could be angled in a way that would reflect an incoming radar beam away from its source, and if the combined shape could still create lift, then a true 'stealth' aircraft was possible.

The irony, of course, was that stealth was in part a Soviet invention. The Soviets had allowed Ufimtsev's paper to be published, because no one east of the Iron Curtain had entertained the idea that such a complex set of equations could ever be applied to an aircraft. It had taken Overholser to pick it out from hundreds of other ponderous Soviet science texts and draw the necessary conclusions.

Stealth, like the bomb, then, had owed its genesis to algebraic formulae. In a sense, it was pure accident that it had been discovered within the nuts-and-bolts world of the aerospace industry.

Podkletnov's apparent discovery of gravity-shielding in Tampere, Finland, had been an accidental by-product of the Russian's work with superconductors.

What they all shared was their origins in pure maths and physics.

That thought gnawed at me as, working through a decade's worth of my files, I began to re-immerse myself in the world of the F-117A Stealth Fighter, Have Blue's operational successor.

Lockheed was used to black programmes. It had built the U-2 spyplane that the CIA had used to overfly the Soviet Union on reconnaissance missions in the late 1950s and it had developed the Mach 3.2 A-12 for the CIA as the U-2's successor. Both projects had come together under rules of draconian secrecy, but stealth, from a security stand-point, would need to be protected even more stringently. The problem facing Rich and his customer in the Pentagon, the Air Force Special Projects Office, was how to go about shielding an entire industry. The U-2 and the A-12 had been built in 'onesies and twosies', but the F-117A was required in multiple squadron strength. Clearly, there would come a time when the programme would need to be revealed, but the longer it could be shielded from public view, the less time the Russians would have to react when it finally emerged into the light.

Rich was one of five Lockheed employees cleared for top-secret work and above. Everyone else connected with the programme – including several thousand factory floor workers – had to be rigorously security-checked. And nothing about their lives was off-limits. 'Security's dragnet poked and prodded into every nook and cranny of our operation,' Rich wrote years later. It almost drove him insane.

As the F-117A began to take shape at Burbank, arrangements were made to accommodate it operationally. A secret base was constructed in the desert at Tonopah, Nevada, and pilots were asked to volunteer without being told anything about the assignment. In New Mexico, at a radar range miles from the nearest public land, the final configuration of the F-117A was subjected to more RCS checks to validate its insect-like radar properties. When the aircraft made its first flight, in June 1981, it did so at Area 51, shielded from the remotest scrutiny by the jagged mountains of southern Nevada and a crack Air Force special forces unit authorised to use 'lethal force' to protect Groom Lake from intruders.

As production ramped up under the administration of the hawkish presidential incumbent, Ronald Reagan, who took office in 1980, it rapidly became clear that the Skunk Works was outgrowing its Burbank facilities and so plans were drawn up to relocate the plant eastwards, to Palmdale, on the edge of California's Mojave Desert. Palmdale was also home to a giant new production facility operated by Northrop, designer of the B-2, the USAF's four-engine strategic Stealth Bomber. The B-2 was black, but not as black as the F-117A. This meant it was acknowledged to exist, even though the rudiments of its design were close-held.

It was in this triangle of land, bordered by New Mexico to the south, California to the west and Nevada to the north-east, that people – ordinary people – started seeing things on moonlit desert nights that they could not correlate with the known facts. Speculation grew that a 'stealth fighter' programme of some description existed – and almost certainly at Lockheed, where security in recent years had tightened considerably –

but the detail was invariably way off the mark. And since the money for the F-117A programme – to the tune of several billion dollars – had been appropriated from the Pentagon's 'black budget', ring-fenced as it was from public scrutiny, that was how things remained until late 1988, when the outgoing Reagan administration revealed just what it had hushed up for so many years.

The F-117A, the bland press release stated, had gone operational in 1983. For five years, its pilots had operated at squadron strength 'in the black', roaming the desert night skies of the US Southwest, practising for the night-time attack mission they would be required to perform in a war. The disclosure had only been made at all because the Air Force wanted to expand the training envelope and start flying the F-117A by day. Otherwise it would have remained buried even longer.

During the F-117A's five years of secret ops, thousands of workers had been involved in the assembly process at Burbank; hundreds more in supporting the aircraft at Tonopah. And yet, not one had breathed a word about it.

The big story in November 1988, then, related to the existence and capabilities of the Stealth Fighter, a programme that rivalled the scale and daring of the Manhattan Project almost 50 years earlier. Like the bomb programme, the F-117A had been meticulously covered up, but unlike the bomb programme, the secrecy had held. With hindsight, this was understandable. The moment the US government had decided not just to develop the bomb, but to use it on the enemy, its security was compromised. Atom spies sympathetic to the Soviet Union – the liberal, heady atmosphere of Los Alamos cultivated them like seeds in a hot-house – were always going to ensure that details of the weapon would be passed to Moscow; and the traffic merely accelerated when it was dropped on the Japanese.

The logic these people had applied in betraying their country was simple. They didn't want America to have a monopoly on such a devastatingly powerful strand of weapons technology. By passing bomb secrets to the Russians, the atom spies believed they were restoring the balance of power and making the world a safer place. It was a lesson that US security managers would not forget.

Stealth was a passive technology. It did not need to be 'dropped' on an unsuspecting enemy, but it had the power to begin and end wars just the same. Because it didn't need to be revealed, it had been easy to cover up, until the moment it suited the US government to make the revelation – to bring it out of the black. Neglected in the process of disclosure, however, was any analysis of the system that had allowed this superhuman feat of engineering and parallel security to materialise; and the black world was a system – a huge, sprawling machine.

Between the bomb and the Stealth Fighter, a massive security structure had swung into place to protect America's ever-proliferating black programmes. And in the vastness of the desert, it wasn't hard for individuals – the little cogs of the machine – to develop a sense of their frailty. The desert had been a good place to build the bomb and the perfect place to field stealth. It had swallowed both programmes – the weapons and the people who'd constructed them – whole.

If anti-gravity was real – and by now I believed the white world contained evidence that *something* was going on – the desert was where it would have come together.

According to Marckus, the black world had known I was coming from the moment I set foot in NASA and had had plenty of time to prepare for my visits.

In London, I struggled to make the right connections.

I returned to something Marckus had said during our meeting by the estuary; something about the Northrop B-2 Stealth Bomber. There was a strand of thought, even amongst some quite high-profile academics, that the B-2 used some kind of anti-gravity drive system. I wasn't inclined to believe the story at face-value. But with scant entry points to the black world, it seemed, for the moment, as good a place as any to start. If there were lessons still to be learned from the media's coverage of the F-117A in the early 1980s – before the programme had been revealed, but when people were reporting things they couldn't explain in the night skies of California, Nevada and New Mexico – the most appropriate was the old one about smoke and fire.

The detour through NASA had pointed up five possible pathways to anti-gravity's application in the black world: manipulating an object's mass and/or inertia; exploitation of the zero-point energy field; perturbations of the space-time continuum; faster-than-light travel; and gravity shielding. Of the five, two (space-time and faster-than-light) were patently out of the question – the B-2 was not a time-machine and it could not even muster the speed of sound – and two more (mass/inertia manipulation and zero-point exploitation) were still some way from a practical breakthrough. Only one, the last of the five, seemed to have resulted in any kind of pay-off. And Podkletnov had tripped over it by accident because of somebody's pipe-smoke, which meant it had been out there, waiting to be discovered, for as long as anyone needed it.

Applied to the B-2, using some of the many billions that had gone into its development, the kind of weight reductions that the Russian had been reporting (anywhere from 2 per cent to 5 per cent) would have been highly significant. The B-2 had already shown its ability to fly half-way round the world and back – a 37-hour, 29,000-nautical-mile return flight from Whiteman Air Force Base, Missouri, to Guam in the Pacific with several air-to-air refuellings en route. With gravity shielding – if such a thing could be moved from the laboratory to a fully-fledged aircraft programme – either its range could be extended or the number of mid-air refuellings cut. Either of these would make the technology a powerful draw.

The biggest argument against the existence of anti-gravity technology had seemed to me to be the lack of any visible hardware from production programmes. Now, I was having to confront the possibility that anti-gravity played a vital part in the most important technological development since the bomb.

Chapter 12

I'd been aware of rumours about the B-2 for some time, but they were so bizarre as to be quite unbelievable, so I'd ignored them completely. In following the thread that had been dropped by Marckus at the estuary, however, I discovered that it led somewhere that was already familiar. The chief proponent of the idea that the B-2 utilised an exotic propulsion source was my old friend Paul LaViolette, the Ph.D who had tripped over the reference to anti-gravity in the Library of Congress.

LaViolette had written an elegant thesis about the B-2 and its anti-gravitational properties soon after the publication of a revelatory article in *Aviation Week*, the 'bible' of the US aerospace industry, in which a number of black world engineers had whistle-blown to the author, the magazine's West Coast editor, Bill Scott, about a host of amazing new technologies supposedly being developed in the deeply classified environment. One of these related to 'electrostatic field-generating techniques' in the B-2's wing leading-edge with a view to reducing the giant Stealth Bomber's radar cross-section.

LaViolette had taken this revelation and coupled it to his own knowledge of Thomas Townsend Brown's work and patents and drawn the conclusion that the B-2 was the embodiment of Project Winterhaven, T.T. Brown's proposal in 1952 to the US military for a Mach 3 saucer-shaped interceptor powered exclusively by electro-gravitics – anti-gravity.

In an essay entitled 'The US Anti-Gravity Squadron', LaViolette outlined the reasoning behind his extraordinarily bold assertion.

The B-2, he held, was simply a scaled-up version of the sub-scale discs, the little charged saucer-shaped condensers, that had whizzed around Brown's laboratories in front of multiple witnesses from the US Air Force and Navy in the late 40s and early 50s. If true, it explained why superficially the US Air Force had shown such little interest in Winterhaven, whilst on the inside – I remembered the recovered transcripts of conversations between air force generals Bertrandias and Craig following the former's visit to Brown's lab in 1952 – those who had seen the tests were going apoplectic with excitement.

While conventional wisdom had it that the B-2's outer skin was composed of a highly classified radar-absorbent material (RAM) that made it invisible to radar, LaViolette plausibly argued that the RAM was in fact a ceramic dielectric material able to store high amounts of electric charge. The material was said to be made from powdered depleted uranium – an incredibly hard substance, commonly used to tip armour-piercing tank shells – with three times the density of the 'high-k' ceramics proposed by Brown in the 1950s. This would give the B-2 'three times the electrogravitic pull' of the Mach 3 saucer at the heart of the Winterhaven study.

Another useful bit of synchronicity was the fact that Brown had, for a short period of time, worked for General Electric, the company responsible for the B-2's engines.

But try as I might, I couldn't get LaViolette's theory to fit. For the thesis to be accurate, it hinged on something approximating a switch in the Stealth Bomber's two-man cockpit that would permit the aircraft to transition from normal jet-powered flight – something, according to La-Violette, that it used on take-off and landing and other occasions where witnesses might be present – to anti-gravity cruise mode.

If there were such a thing as this switch, I knew, pound to a penny, dollar to a cent, its existence would have leaked.

The B-2 first flew in 1989 and the first operational aircraft arrived at Whiteman Air Force Base in 1993. A total of 20 aircraft had been built at $2 billion a shot and all of them had been delivered by the end of the decade. Dozens of crewmembers had been trained in that time, and many of them had been interviewed. The B-2 had started as a black programme (or more accurately, a dark grey one, since its existence had been confirmed officially from the start, all other details being classified). But now it was in the open, albeit, still, with some highly classified features. Pilots – an inherently talkative breed – would have let it slip somewhere along the line. Rumours would have developed about such a switch. But they never had. Moreover, B-2 pilots and engineers hadn't just denied the anti-gravity story, they had scoffed at it openly. Intuitively, my colleagues and I were agreed, the B-2 anti-gravity story was wrong. In fact, the whole notion of it was absurd.

The end result was that 'the B-2 as anti-gravity vehicle' had been consigned to the columns of conspiracy mags and tabloids; the mainstream aerospace press, afraid of an adverse reaction from its conservative readership, had refused to touch it.

With one exception.

Enter Britain's most eminent aerospace journalist, Bill Gunston, OBE,

Fellow of the Royal Aeronautical Society, and an article he'd penned entitled 'Military Power'.

Gunston, who served as a pilot in the RAF from 1943 to 1948, was scrupulous over his facts – as editor of Jane's yearbook on aero-engine propulsion, he had to be. 'Military Power', published in *Air International* magazine, was a walk-through dissertation on the development of aero-engine technology since the end of the Second World War; good, solid Gunston stuff.

Until the last couple of pages, when, to the uninitiated, it appeared that the aerospace doyen had lost his mind. Gunston not only portrayed the B-2 anti-gravity drive story as fact, but went on to reveal how he had been well acquainted with the rudiments of T.T. Brown's theories for years, but had 'no wish to reside in The Tower [of London], so had refrained from discussing clever aeroplanes with leading edges charged to millions of volts positive and trailing edges to millions of volts negative.'

Gunston explained why he felt that there was much more to the B-2 than met the eye, drawing on a lifetime of specialised knowledge. In short, if you applied the laws of aerodynamics and basic maths to the known specifications of the B-2, there was a glaring mismatch in its published performance figures.

It was clear, Gunston said, that the thrust of each GE engine was insufficient to lift the 376,500 pound listed gross weight of the aircraft at take-off.

The only way the B-2 could get into the air, therefore, was for it somewhere along the line to shed some of its weight. And that, of course, was impossible, unless you applied the heretical principles of Thomas Townsend Brown to the aircraft's design spec. The B-2 wasn't a black world aircraft programme anymore. It was operational, had been blooded in combat and was frequently exhibited to the public. If it used an anti-gravity drive system, the proof of the phenomenon existed in plain sight.

If it didn't, then someone, somewhere had gone to some lengths to make a number of highly respected aerospace experts believe that it did.

In October 1990, *Aviation Week* published a report based on an analysis of 45 different daylight and night-time eyewitness observations of strange aircraft over the desert Southwest – Bill Scott's backyard. In this, he concluded that there were 'at least two – but probably more – types of vehicles' beyond the F-117A and B-2 under test. One was a 'triangular-shaped, quiet aircraft seen with a flight of F-117A stealth fighters several times since the summer of 1989'; another was a 'high-speed aircraft characterised by a very deep, rumbling roar reminiscent of heavy lift rockets'; a third was a 'high-altitude aircraft that crosses the night sky at extremely high speed . . . observed as a single, bright light – sometimes pulsating – flying at speeds far exceeding other aircraft in the area.' Because the latter was always seen above 50,000 ft, where you wouldn't routinely expect to hear engine noise, Scott deduced that the second and third aircraft might be one and the same type.

Speculation that the Reagan administration had embarked on a massive programme of black world aerospace and defence research began circulating in earnest amongst Washington reporters and defence journalists following the publication of a 'line item' called 'Aurora' in the Pentagon's fiscal year 1986 budget. Aurora was supposedly a 'sleeper site' used by the Pentagon for burying B-2 funds at a time when the Stealth Bomber was still highly classified. But this explanation, put about by 'highly placed Pentagon sources' when the Aurora story hit the streets, simply smacked of back-pedalling. As Aurora was listed under the 'Other Aircraft' category and was projected to receive a huge increase in funding between fiscal years 86 and 87 – a jump from $80.1 million to $2.272 billion – the sleuths concluded that its inclusion in the budget request was a genuine mistake, that the aircraft was real and, because it was listed in the 'reconnaissance' section, must have been a replacement for the Mach 3.2 Lockheed Blackbird, which was earmarked to retire from its strategic spying duties in the early 1990s.

Then, when that dark triangular shape was seen by a trained Royal Observer Corps spotter over the oil rig in 1989, people put two and two together and deduced that Aurora wasn't just a line-item, but a real flying aircraft.

One of the supreme ironies of the black budget system is that, thanks to US bureaucratic procedure, you can work out, almost to the cent, what the Pentagon is spending on its most deeply classified programmes. Every year its financial planners submit an unclassified version of the defence budget to Congress. It is a remarkably detailed document, listing major line items of expenditure in the military research and development (R&D) and weapons procurement arenas. Trawl down the list, however, and some of the dollar values of these line items are missing. Many of those in this category have strange-sounding codenames like Forest Green, Senior Year, Chalk Eagle or Centennial. Others have curiously nondescript headings such as 'special applications program' or 'special analysis activities'.

These are black programmes. Because the unclassified budget includes accurate total budgets for all three armed services, all you need do is subtract the total dollar value of the line items from each category, and the difference is what the Pentagon is spending on black world research and acquisition.

In 1988, the total was computed to be $30 billion for R&D and secret weapons programmes – more than the entire annual defence budget of a major European NATO nation such as Britain, France or Germany.

What we were talking here was an industry within an industry, multiple layers deep and compartmentalised to death horizontally. The black world's inaccessibility was based not just on its segregation from the outside, but on the inability of managers and engineers in one of its compartments, or on one of its layers, to have oversight of anything that wasn't within his or her 'need to know'.

In a system whose existence had been designed to be denied, examples were few and far between, but one shone clear. General George Sylvester, head of the Air Force's R&D programme in 1977, was not 'accessed' to the Lockheed Have Blue programme, even though the programme itself was run out of his office at Wright-Patterson Air Force Base. One of the 'bright young colonels' at Air Force Special Projects in Washington who ran the Air Force's portion of the black budget had decided to cut Sylvester out of the loop.

It was small wonder, then, that reporters and investigators began to level the charge that the black world had spiralled out of control.

How deep did the layers go? How many compartments did the system contain? What kind of technologies were being pursued within it?

The truth was, nobody knew. The black world was labyrinthine and it was essentially unaccountable; its whole system of compartmentalisation so convoluted, so tortuous, that if people like Sylvester had been excluded, then God only knew how many others who should have had the 'need to know', didn't.

It was in this climate of hidden expenditure that Bill Scott reported what he was seeing and hearing in and around Palmdale.

The prognosis was that Aurora was a massive leap forward in aerospace terms, that it was powered by a new form of 'combined-cycle' engine fuelled by liquid methane or hydrogen to give it a cruise speed anywhere up to Mach 8 – 5,300 mph. It was also a fair bet, Scott and others reported, that the Lockheed Skunk Works, which was in the process of shifting to Palmdale, was the firm responsible for building it.

Outwardly, the Skunk Works had next to no work on the stocks, but something was keeping its 4,000 employees busy just the same and something was rattling the windows of the desert Southwest at night.

From mid-1991 until the mid-90s, the US Geological Survey, the agency that monitors earthquake activity, revealed that an unknown aircraft, with sound footprints different from the SR-71 or the Space Shuttle, was causing 'airquakes' as it crossed the California coastline from the Pacific, heading inland for southern Nevada. Couple this with some fresh US eyewitness sightings after the North Sea encounter of a huge, arrow-shaped aircraft that left a weird, knotted contrail in its wake and made a noise 'like the sky tearing apart', and a recent programme of expansion at the Air Force's test site at Area 51 in southern Nevada, and it all added up to a black world economy that was, so to speak, booming, too; a situation that continued even after the Cold War ended.

Following the dissolution of the Soviet Union, the US Air Force budget, in concert with those of the other armed services, was cut and black world funding along with it. But as a proportion of the overall budget, the classified segment of the USAF's R&D and acquisition spend actually increased and, even in the mid-late 1990s, hovered comfortably at around the $11 billion a year mark. This left more than enough money for the Air Force to pursue a range of exotic technologies for the 21st century.

In early 1992, Scott was contacted by a group of black world engineers who had decided to break cover and talk about some of the programmes they were working on. Their reasoning was simple. The Cold War was over and there was no need for the US government to protect some of its most highly classified technologies any longer. Some of these technologies, the engineers told the reporter, could actually be spun into the white world of commercial aerospace programmes to give the US an enormous economic advantage over its international aerospace trade rivals.

In revealing their presence to Scott, they acknowledged that they were breaking a 'code of silence that rivals the Mafia's'. The dangers of their approach were already manifest. Two engineers said that their civil rights had been 'blatantly abused' by their black world controllers in a bid to prevent them from leaving the deeply classified environment.

'Once you're in, they don't let you go,' one of them said.

Amidst the revelations that the engineers dangled in front of Scott were a number of technologies for reducing the visible signatures of the B-2.

To people in the business, Northrop's winning of the $40 billion Stealth Bomber competition against Lockheed in 1981 had long been something of a mystery. It was, after all, Lockheed that had made the breakthrough in mathematically computing stealth in 1975. But then, the F-117A, composed of its mass of angled flat panels, was 'first-generation stealth'.

The B-2, or the Advanced Strategic Penetrating Aircraft, as it was known when it was still deeply grey, was a second-generation aircraft and this, according to Northrop, was manifest in its blended aerodynamic shape. Instead of being rough and angular like the F-117A, it was smooth and rounded – a product, Northrop claimed, of five years' worth of improvements in computing technology between the start of Have Blue programme and the ASPA.

Improved software algorithms processing had allowed Northrop to come up with a shape – a refinement of the YB-49 flying wing jet bomber it had first flown in 1947 – that allowed the aircraft to look and fly right *and* to remain invisible on enemy radar screens.

Not only that, but according to LaViolette it had the capability to fly electrogravitically as well.

In my basement, where the noise of the city couldn't intrude upon the sort of silence you needed to sift a decade's worth of documents and files on the arcane properties of stealth, I still wasn't happy about LaViolette's assertion, because of the absence of that all-important switch, the one that kicked in the B-2's anti-gravity drive system.

There was one other beguiling aspect to Scott's story. His sources said that charging the airframe electrostatically not only helped to make the B-2 stealthy, it also reduced friction heating of the airframe, its sonic footprint and its drag.

This last point was crucial. If it were possible to alter the 'drag' or air resistance of an aircraft, you could either make it fly further or faster or both.

The drag reduction system was supposed to work as follows. By creating an electrostatic field ahead of an aircraft, it ought in theory to repel air molecules in the aircraft's path, allowing the plane, in effect, to slip through the atmosphere like a thin sliver of soap through warm water.

Like anti-gravity itself, the whole notion of using electricity fields to reduce aircraft drag was heresy – it was, according to traditionalists, just smoke and mirrors. But here, in Scott's article, was a reference to Northrop having conducted high-grade research in 1968 on just such a phenomenon: the creation of 'electrical forces to condition the air flowing around an aircraft at supersonic speeds'. Furthermore, the California-based aircraft company had supposedly submitted a paper on the subject to the American Institute of Aeronautics and Astronautics that same year.

I hit the Net and went into the AIAA website, then did a title search. But the papers on the AIAA site only went back as far as 1992 and I was out of time.

I rubbed my eyes. In two days I was due to attend an aerospace convention in Las Vegas. And I wasn't any closer to solving the puzzle.

Or was I?

During the Second World War, I remembered Thomas Townsend Brown had been involved in experiments that sought to show how you could make a ship disappear on a radar screen by pumping it with large doses of electricity. Between 1941 and 1943, Brown had supposedly been involved in tests, I saw now, *that were identical in principle to the methodology that Northrop seemed to have applied to the B-2 to make it the ultimate word in stealth*. Researchers had never taken Brown's wartime experiments seriously because the precise nature of the work had been obscured by the myth of parallel dimensions aboard the USS *Eldridge* – the ship at the heart of the 'Philadelphia Experiment'. Brown had also used electrostatics to power his model discs and, by his own testimony and that of witnesses – including a general from the US Air Force – they had flown by defying gravity.

Could it really be that in researching the B-2 I had picked up the threads of T. T. Brown's work again? Work that the US had continued to develop in the black?

I jotted the keywords on a piece of paper, like a crossword player consigning letters to a page in the hope of unravelling an anagram. *Brown – electrostatics – stealth – anti-gravity* – and then, against my better judgment – *parallel dimension*.

I studied the words for a long time, but nothing jumped at me.

Below: Alleged photograph Rudolph Schriever.

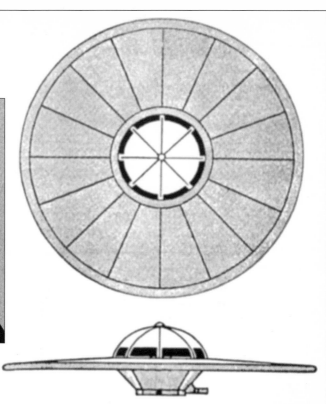

Right: The Legend's rendition of the Schriever/Habermohl/Miethe disc.

Right: Richard Miethe.

Below: Artist's impression of foo-fighters engaging a formation of B-17 bombers. (All images via Bill Rose)

Above: Thomas Townsend Brown.

Right: Patent illustration showing experimental layout of the charged condenser plates Brown demonstrated to the US Air Force and Navy.

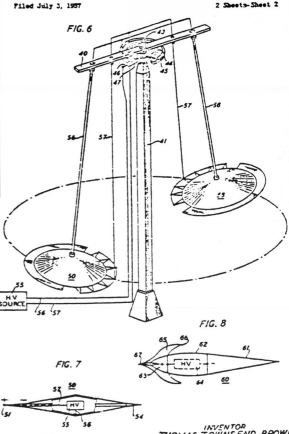

Aug. 16, 1960

T. T. BROWN

2,949,550

ELECTROKINETIC APPARATUS

Filed July 3, 1957

2 Sheets-Sheet 2

FIG. 6

FIG. 7

FIG. 8

INVENTOR
THOMAS TOWNSEND BROWN
BY
ATTORNEYS

THOMAS TOWNSEND BROWN: BAHNSON LAB 1958-1960

ELECTROGRAVITICS LEVITATION EXPERIMENTS

ORIGINAL SILENT FOOTAGE TRANSFERRED TO VHS VIDEO

See New Styles of Electrogravitic Prototypes Designed for Vertical Flight and Hovering, Captured on Film at the Bahnson Laboratory

Above: Brown holding a disc capacitor prior to charging.

Left: A demonstration of electrogravitic lift at the Bahnson Laboratories in the late 1950s.
(All images via the Integrity Research Institute)

Left: John Frost.

Below: A cutaway illustration of the Avro Silverbug showing the radial flow gas-turbine.

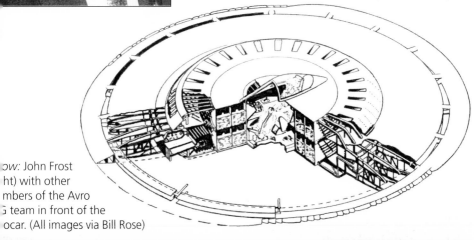

...ow: John Frost
...ht) with other
...mbers of the Avro
...i team in front of the
...ocar. (All images via Bill Rose)

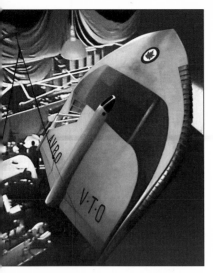

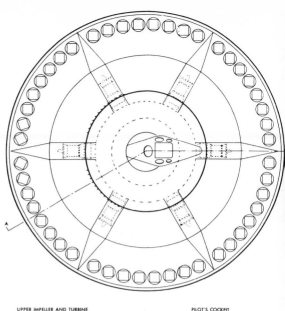

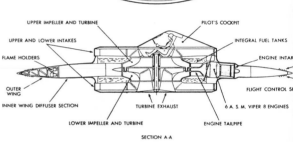

Above: A full-scale engineering mock-up of the Project Y Manta at Avro's factory in Malton, Ontario.

Right: General arrangement of the proposed Avro Project 1794, showing arrangement for its six Viper turbojets.

Below: Avro's test-rig for the Mach 4 Project 1794 saucer. (All images via Bill Rose)

UPPER IMPELLER AND TURBINE
PILOT'S COCKPIT
UPPER AND LOWER INTAKES
INTEGRAL FUEL TANKS
FLAME HOLDERS
ENGINE INTAK
OUTER WING
FLIGHT CONTROL SH
INNER WING DIFFUSER SECTION
TURBINE EXHAUST
6 A.S.M. VIPER 8 ENGINES
LOWER IMPELLER AND TURBINE
ENGINE TAILPIPE

SECTION A·A

Above: British Aerospace's concept of an advanced fighter interceptor, powered by an anti-gravity t engine.

Below: Another British Aerospace design for a heavy-lift anti-gravity vehicle. (Both images via BAE stems)

Above left: Tom Valone of the Integrity Research Institute. (Tom Valone)

Above right: Paul LaViolette. (Paul LaViolette Ph.D.)

Below: Dr Evgeny Podkletnov (left) with Dr Ron Evans, Head of BAE Systems' Project Greenglow. (BAE Systems)

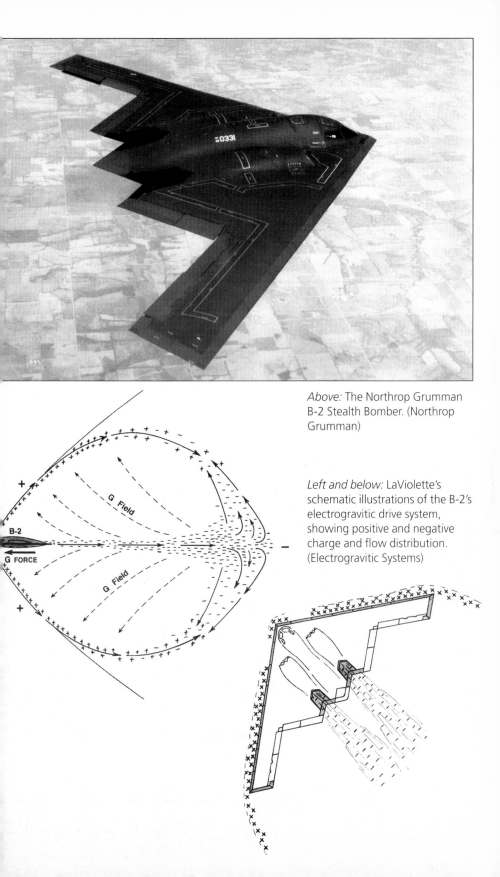

Above: The Northrop Grumman B-2 Stealth Bomber. (Northrop Grumman)

Left and below: LaViolette's schematic illustrations of the B-2's electrogravitic drive system, showing positive and negative charge and flow distribution. (Electrogravitic Systems)

Left: The Lockheed Martin F-117A Stealth Fighter over typical desert terrain.

Below: 50 years of Skunk Works Products – those that the company has owned up to.

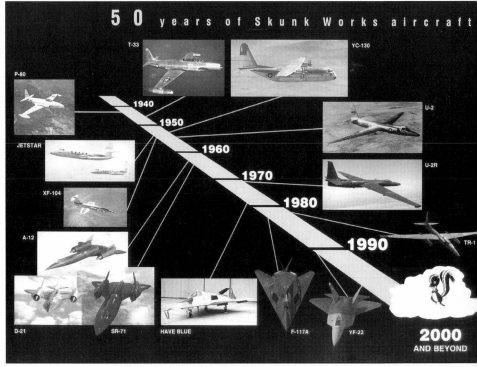

5 0 years of Skunk Works aircraft

T-33
YC-130
P-80
1940
1950
1960
U-2
JETSTAR
1970
U-2R
XF-104
1980
A-12
1990
TR-1
D-21
SR-71
HAVE BLUE
F-117A
YF-22
2000
AND BEYOND

Right: Ben Rich and the F-117A. (All images via Lockheed Martin)

Above: Map showing key California-based sites of the Skunk Works in the late 80s/early 90s. (Lockheed Martin Skunk Works)

Right: Jack Gordon, head of the Skunk Works in 1996. (Lockheed Martin Skunk Works)

Below: An artist's rendering of Aurora, the USAF's mythical hypersonic spyplane. (Julian Cook)

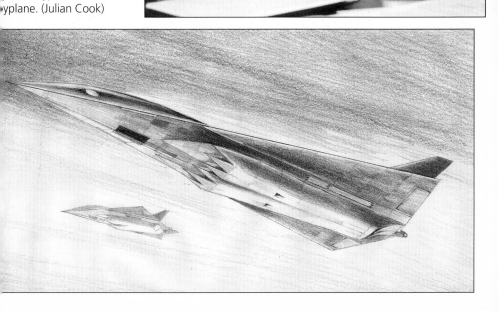

Below: SS General Hans Kammler – the only known photograph of him in uniform; France, 1944.

Above: A partially lined gallery at the SS-run 'Giant' underground manufacturing complex near Waldenburg, Poland. (M. Banas)

Left: A completed gallery at the 'Giant' underground complex. (M. Banas)

bove: Viktor Schauberger and his implosion-based home power generator in 1955, with interior
ot.

Left and below: The Schauberger Repulsine with and without cone-shaped intake. (All images via Schauberger Archive)

Left: Bob Widmer, chief designer of General Dynamics Convair, in his office in the late 1950s. (Lockheed Martin)

Below: A grainy shot of a wind-tunnel model of Kingfish, Widmer's Mach 6 spyplane proposal. (Air International)

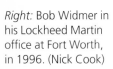

Right: Bob Widmer in his Lockheed Martin office at Fort Worth, in 1996. (Nick Cook)

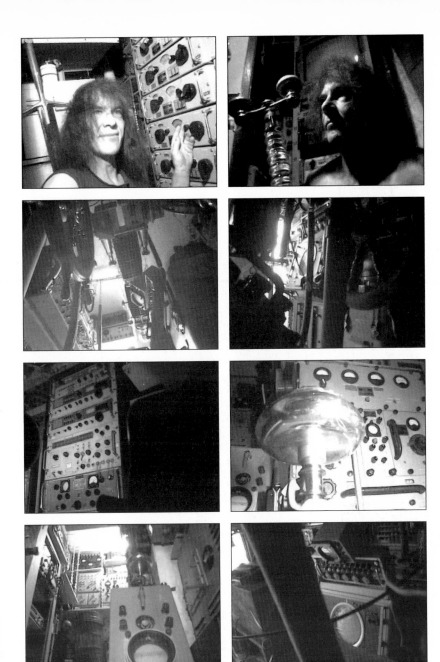

John Hutchison in his Vancouver apartment, surrounded by equipment integral to the Hutchison Effect. (John Hutchison)

Metal ingots 'disrupted' by the
Hutchison Effect. (John Hutchison)

Chapter 13

Vegas. Two thumps through the 747 and I was on the ground at McCarran International, an airport, like Miami's, that crackled with illicit activity; you could just feel it. We taxied to the terminal and passed engineering shops filled with propeller planes that looked like they should have been scrapped decades ago; old classics running freight operations now, with radial engines that dripped oil on the ramp and coughed clouds of smoke, their tired but sturdy airframes lined with irregular rows of rivets and bumps.

I raised my eyes and took in the view. Vegas, sprawled there just beyond the perimeter fence under an upturned bowl of blue sky, half a dozen giant new hotels since I'd been here last, half a dozen more going up amidst a thicket of cranes.

I loved and loathed this city with equal measure.

Down on the Strip, with the sun setting on this city of broken dreams, I took in the glare of the lights from the back of the cab, half-listening to the driver's lament over the mob's exodus and its replacement by corporate America.

The slipping sun played over the tops of the mountains that bordered the southern end of the Nellis Air Force Range, a closed-off piece of government land the size of Switzerland where the Air Force tested its latest military hardware and ran air warfare exercises, and the driver's voice still filling my head about the days when people in the city had lived and died by a set of rules everyone understood.

I grunted something neutral as I tracked the Convention Center where tomorrow I'd tramp the hallways, doing what I always did on these occasions: talk to old friends, meet contacts, trawl the stands, hoover up data, collate stories, file copy, hit the town, go to bed late, get up early and then go through it all over again.

The cab pulled into the Sahara, a hotel that exuded the Vegas of Sinatra, Martin and the mob; the mid-50s, when it had opened.

It had been facelifted so many times, it seemed like the demolition teams would be here at any moment, but through half-closed eyes you

131

could still see how it once had been, which gave it a strange kind of charm. The Sahara suited me just fine.

Its pile-em-in, rack-em and stack-em ethos, ideal for gamblers passing through, meant it was shunned by most of the show delegates, but it had the enormous advantage of being a stone's throw from the Convention Center.

The following day, I picked up my reporter's pass and filed into the exhibition hall. It was jam-packed with aerospace and defence companies from across America, their stands and booths ranged in rows either side of the carpeted walkways.

From all over the hall, TVs plugged into videos on permanent loop flashed with images of combat aircraft going through their paces or weapons under test, the pictures backed by dead-pan monologues and Top Gun-style soundtracks.

I slipped into the routine, walking the course, checking the stands for signs of developments or breakthroughs.

Sometimes stories came from a new aircraft, spacecraft or weapon system depicted in model form or artist's impression, but such copy as this generated was merely the low-hanging fruit – things that the public affairs people wanted you to see. As long as it constituted news, most reporters, myself included, were happy to go through the motions.

More often than not, though, the developments we looked for were more subtle: a missile fitted with a new seeker or an aircraft with a new sensor, denoted by an aerial or antenna-window on the display model that hadn't been there last time around.

Such changes, though seemingly inconsequential to an outsider, pointed to performance gains that signalled real news to a defence reporter, news that the industry's primary customers generally didn't want to see in print.

But there was another side to show reporting, a side that I altogether preferred: chewing the fat with people in the business I'd known and liked for years, gathering data that gave hints of longer-term trends; what was really happening to the industry in the face of Cold War budget cuts, where the technology was headed, who was buying, who wasn't.

Three days at an aerospace and defence exhibition gave you information you couldn't get from a month's worth of phone calls. And after the intensity of weeks spent scrutinising obscure pieces of data for hints of a black world anti-gravity programme, the change was refreshingly good.

On the third day, long after the press conferences had dried up and I'd filed my final piece of copy to the magazine, I hit the Northrop Grumman stand and started talking to a guy who'd worked a whole range of the company's programmes, including the B-2 Stealth Bomber.

Careful not to provoke the reaction that usually greeted the anti-gravity question, I asked him what he made of the stories that the B-2 charged its surface electrostatically.

This time, no derogatory smirk, just a straight answer and a half-smile.

It was impossible, he told me. If you charged the airframe of the B-2, you'd fry its on-board electronics. He went on to describe a secondary problem. All aircraft need to discharge the static electricity they absorb naturally as they move through the air – usually via small wire aerials on the wing-tips. Deliberately adding to this static build-up with a man-made electrical charge would be enormously risky, the engineer maintained. Since it would be impossible to discharge the electricity fast enough, it would turn the aircraft into a giant lightning conductor.

I called Marckus from my hotel room the following morning. The eight-hour time difference meant that he was at home, the day's work finished. Classical music filled the background, Brahms's Symphony Number Four turned deafeningly high, giving me, in a moment of insight, a clear view of Marckus' domestic situation. I knew then the man lived alone.

'He said what?' he shouted back.

I repeated the engineer's assertion. Marckus left the phone for a few seconds to turn the music down.

'The man's an idiot,' he said when he came back. 'Either that or he deliberately lied to you.'

'Why should he do that?'

'Why do people normally lie?'

'Come on, Dan, it's too damned early here for twenty questions.'

'The guts of the B-2 are held within a Faraday Cage,' Marckus explained. 'It doesn't matter how much electricity you pump across the skin. The inside remains insulated from it. The guy was bullshitting you. And it's not even very good bullshit.'

The rest of his argument was badly flawed also, Marckus said. If you looked at the B-2 from the top, it looked like a boomerang, with straight, sharp leading edges and a serrated trailing edge. Regular aircraft had two wing-tips; two points from which to discharge their natural build-up of static. But the B-2 was a flying wing. With its saw-tooth back-end, it had, in effect, seven wing-tips from which to discharge the continuous flow of electricity from its supposed flame-jet generators.

At low altitude and low speed, the discharges would be difficult, if not impossible to see with the naked eye. But at cruise altitude, with enough plasma build-up around its leading edges, an aircraft – particularly a supersonic or a hypersonic platform – would glow like a light-bulb. It would pulse.

A high-altitude aircraft that crosses the night sky at extremely high speed . . . observed as a single, bright light – sometimes pulsating – flying at speeds far exceeding other aircraft in the area.

One of the night-time eye-witness observations reported in the October 1990 edition of *Aviation Week*.

'Northrop has been playing around with electrostatics for years,' Marckus said. 'In 1968, it submitted a paper called "Electroaerodynamics in Supersonic Flow" to the American Institute of Aeronautics and Astronautics.'

'I know,' I told him, happy for once I had something he didn't. 'I tried to find it on the AIAA website, but the papers on their electronic database don't go back far enough. I'll have to request a copy from their archives instead.'

Even across eight time-zones, I caught Marckus' withering sigh.

'Don't bother. I applied eight years ago. It's missing. They couldn't find any record of it. Almost certainly, some special projects office in the Pentagon pulled it. Made it disappear. That happens occasionally. Someone fucks up and the black world has to go clean up. Actually, it wasn't all Northrop's fault. Back in the 60s, they had some noble ideas that this technology could be applied as a drag-reduction device for supersonic airliners, cutting heat friction and fuel-burn. It was only when someone clocked the fact that a plasma shield around an aircraft also reduces its radar signature that the military really woke up to it. It's been rumoured for a while that Northrop had come up with something revolutionary in this field. The missing paper proves it.'

Marckus carried on. I closed my eyes and tried to think.

The signal I was picking up was painfully indistinct against the background mush of interference, but I could see that it added up to something important.

The B-2 clearly used some kind of skin-charging mechanism to lower its radar (and quite possibly its heat) signature. But it was also clear, from the nature of the paper that Northrop had written in 1968, that the origins of this electric shield, this stealth cloaking device, were actually rooted in another field of research altogether – drag reduction. According to Marckus, who now admitted to having witnessed similar experiments in the UK, the electrostatic field could also be made to behave like a big 'jet-flap', an extension of the wing. Deploy this invisible jet-flap on take-off and it was the same as if the wing area had been substantially increased, with a parallel reduction in drag and a concomitant improvement in lift – by perhaps as much as 30 per cent.

Bingo.

This could explain how the B-2, or an aircraft like it, staggered into the air with a full bomb and fuel load. Thirty per cent seemed far too high, but even a few percentage points would constitute a breakthrough. A substantial lift improvement was the equivalent of a substantial reduction in weight. Maybe the B-2 lost weight, so to speak, aerodynamically, not electrogravitically. It would explain why Northrop had beaten Lockheed to the B-2 contract. While the Skunk Works had been playing around with Maxwell's equations and two-dimensional shapes, Northrop had already worked out how an aircraft could be enveloped in a shield of static electricity and how that energy could be put to work: as both a cloaking device and a method for reducing its air resistance.

No laws of physics broken, no anti-gravity switch in the cockpit, but an extraordinary development nonetheless.

A weight reduction device by proxy.

And in working this through, I realised something else. The B-2 anti-gravity story actually played into the hands of black world security. It was so fundamentally unbelievable that it discredited any serious investigation of the application of electrostatics to aircraft. What defeated me, was why the spooks should have worked so hard to stop its disclosure – pulling papers from the AIAA and so forth – and yet be content for details to be released of the Lockheed stealth method.

But then, disinformation works best when you mix it in with a little truth.

It was then that the picture cleared momentarily and I glimpsed something through the snow and distortion. The keywords I'd written down and then hurled into the bin in the basement of my home a few days earlier swam into focus. If I could just hold the picture a second or two longer . . .

Marckus' voice, punchy and impatient on the other end of the line, brought me back to the here and now of my room in the Sahara.

I'm still here, I told him.

He was asking me where I was headed next.

I told him there was only one possible place I could go. If Northrop had been working on electrostatics since the 1960s, *if* it was now operationally deployed on a $2 billion front-line bomber that had been designed in 1980, there had been plenty of time in the intervening period for improvements and refinements. I knew there was no way that Lockheed, now Lockheed Martin, would have taken Northrop's B-2 victory standing still.

Aviation Week had written about the 'fast mover' that had been observed crossing the night sky as a single, pulsating point of light. And others had reported the manifest contradiction of a prominent Air Force contractor at Palmdale that was filled to capacity with workers, was hugely profitable, yet had nothing to show for it.

I told Marckus that my request to interview the head of the Skunk Works, Jack Gordon, had just been given the green light.

Better still, it was to take place within the holy of holies itself.

It was only on the road to Lancaster, my next stop, across the border in California's Antelope Valley and a three-hour drive across the Mojave Desert, that I found the quiet I needed to work through what I'd been unable to finish in my call to Marckus.

Its starting point was one of the keywords that I'd written down on the note I'd penned and then trashed five days earlier: stealth. It all started and ended with stealth. In 1943, Brown had been involved in an experiment that sought to make a ship radar-invisible. His test was the first attempt by anyone – certainly, as far as I knew – to have conducted a genuine radar-stealth experiment.

At its heart was a technique developed by Brown to wrap a ship in a field of intense electromagnetic energy, produced by powerful electrical generators.

In the mid-1970s, the US Department of Defense was briefed by Lockheed on a new form of technology – essentially, to do with airframe shaping – that reduced the radar cross-section of a fighter-sized aircraft to that of a large insect.

In 1976, the company was awarded a contract to develop the Have Blue stealth demonstrator. Two years later, Lockheed was given the go-ahead to develop the F-117A, the world's first 'stealth fighter'.

A new military science had been born – one that was about to change the defence planning of entire nations.

As a result, the technology was classified to the rooftops. Stealth had become America's biggest military secret since the bomb.

It was then that somebody must have remembered Northrop's research papers into drag-reducing plasmas and dusted them off, recalling, probably, that electrostatics had a highly beneficial spin-off effect: it could reduce – perhaps even cancel altogether – the radar signature of an aircraft.

A picture formed in my mind's eye of an insect-sized image on a radar-screen winking out altogether. The B-2, three times the size of the F-117, would be 100 per cent radar invisible if it worked as advertised with its plasma cloak activated.

Since the development of big-budget weapon systems is all about risk-reduction, the Northrop approach, fundamentally different from Lockheed's, would have appealed to the bright young colonels of Air Force Special Projects.

If one of the methods failed to deliver, you could always rely on the other one to show you the money. It gave the colonels two bites at the problem.

Soon afterwards, Northrop won the contract to build the B-2A Stealth Bomber, based on its use of electrostatics to make it ultra-stealthy. The technique, first developed by Brown, also improved its aerodynamic efficiency and reduced its drag. It represented a phenomenal breakthrough – a development that would give the US a ten-year lead, maybe more, over the Soviet Union.

But herein lay a problem.

The whole science of stealth would need to be protected like no other strand of weapons science before – the lessons of the bomb being uppermost in the minds of those charged with securing the Air Force's precious new secret.

A novel protection strategy, one that overcame the flawed security model established to protect the bomb 40 years earlier, needed to be set in place.

Eradication and disinformation, coupled with the secure compartmentalisation of the programme, formed the basis of that strategy.

From that moment on, past clues or pointers to stealth would have slowly and carefully been excised from officialdom. It was most likely at or around this point that the Northrop paper was pulled from the archives of the AIAA.

But then, in the archive-trawl, somebody would have come across the work of T. T. Brown. Brown's stealth work for the US Navy in the early 1940s, coupled with his use of electrostatics in his electrogravitic work, was an unexpected discovery that needed to be handled with great care. The man's work was a dangerous liability. Something needed to be done – something truly innovative.

In 1978, the authors of the Philadelphia Experiment, Berlitz and Moore, were apparently approached by two US Air Force officers who claimed to have knowledge of a fantastic secret. They told how they had had a chance encounter with a man who'd claimed to have been on the USS *Eldridge* in 1943.

The man had told the two officers how the warship had vanished from its moorings during a test that sought to make it invisible to radar.

In 1979, after further research by Berlitz and Moore, much of it supplied by their elusive and slippery source Carlos Allende, aka Carl Allen, *The Philadelphia Experiment* appeared in print – and Brown's name became forever linked with the myth of the *Eldridge*.

The story's primary twist was that the ship had been transported into another dimension – kooky stuff and, to any sensibly minded individual, wholly unbelievable.

In the 1980s, the decade when stealth came to be fielded operationally, the technology had acquired its own tailor-made cloak of disinformation.

Tempting as it was to assume that Winterhaven or something like it had been pursued in the black since the 1950s, there was simply no evidence for it.

There was no doubt in my mind that LaViolette's articles on the B-2 were sincere, but they had only helped to muddy the Brown story further. When the B-2 and its anti-gravity propulsion system appeared in print, Brown – whose work was at the heart of this technology – was already a highly discredited figure. Whichever way you approached the story of his life and work, sooner or later you hit the Philadelphia myth. It blew all hope of dispassionate reporting on the man out of the water.

It turned out that the removal of the AIAA paper wasn't the only example of censorship in this field. Before leaving Vegas, I'd called Tom Valone of the Integrity Research Institute. Valone told me that he had been trying for years to locate a copy of an appraisal of T.T. Brown's experiments conducted in 1952 by the Naval Research Lab in Washington D.C., but had been informed that the NRL no longer had it. *The T.T. Brown Electrogravity Device: A Comprehensive Evaluation by the Office of Naval Research* had disappeared into the ether as well.

To deduce from this that Brown had cracked gravity was too big a leap for me. In his experiments, he had demonstrated to the satisfaction of many that an anti-gravity effect was possible. But that didn't mean he'd got to the point where the technology could power an aircraft. Project Winterhaven had promised to deliver this, but Winterhaven never got off the ground. Not in the white world, anyhow.

Somewhere between the junction at Barstow and Boron, a tiny mining community on the edge of the Mojave, the sun finally slipped behind the southern tip of the Sierra Nevadas and I knew I'd taken the disinformation angle as far as it would go. There were just two things that wouldn't fall into place. One of them related to the words on that self-penned note. Brown, electrostatics, stealth and anti-gravity: all of these elements were bound together in the disinformation story – those who had assembled it operating on the age-old principle that a little truth really does go a long, long way.

Why, then, did they introduce the fifth element? The part about the *Eldridge* slipping into a parallel dimension? How had such an improbable notion popped into the heads of the disinformation gurus?

Somehow, I had the feeling Marckus knew about all of this, yet hadn't passed the knowledge on. As I hit the freeway and turned west towards Lancaster, night falling as fast as rain, it was this thought more than any other that troubled me most.

The Desert Inn motel, its lobby replete with signed photographs of test pilots, display-case models of their aircraft and other Right Stuff memorabilia, was located on an invisible confluence of routes connecting the various nodes of the Air Force's classified desert kingdom. Twenty-five miles to the north lay Edwards Air Force base, home of the USAF's premier test pilots from the days when Chuck Yeager had broken the sound barrier in the Bell X-1. Proof of a link between Edwards and Area 51, the Nevada test base that didn't officially exist, 250 miles away as the crow flew, had been ascertained in a neat piece of detective work by Glenn Campbell, the leader of an outfit called the Desert Rats, a group of dedicated 'stealth-watchers' who hung out in the scrubland around Area 51, hoping for a glimpse of some exotic piece of classified hardware. Campbell had come by a procedures manual for security personnel assigned to Area 51 in which they were advised to tell anyone who professed any interest in them that they were attached to a facility called 'Pittman Station'. Going through local county records, Campbell found papers attesting to the fact that these personnel had been sworn in as local deputies. One of them had listed his mailing address as 'Pittman Station, Henderson 89044', but all of them said that they were affiliated to something called 'DET AFFTC'. Campbell knew that AFFTC stood for the Air Force Flight Test Center at Edwards.

A powerful Web search engine did the rest. It turned up a 1990 NASA résumé for an Air Force astronaut candidate called Captain Carl Walz who was employed at 'Pittman Station, Nevada', which in reality, Campbell knew, was nothing more than an abandoned electricity substation on the Boulder Highway. Another biography, found by a Net sleuth colleague of his, ascertained that Walz had served as a flight-test programme manager at 'Detachment 3 of the Air Force Flight Test Center at Edwards'. Because there was no Det 3 at Edwards itself, Campbell deduced that it had to be Area 51 and that Pittman Station was the mail drop.

Edwards, then, was the white world mirror-image of Area 51, the super-classified facility at Groom Lake.

Both the 'pulser' and the aircraft tracked by the US Government Survey's seismic equipment (more than likely one and the same aircraft type) were viewed on a heading that either led into or out of Groom Lake.

Twenty miles to the west, somewhere behind the palms surrounding the Desert Inn's swimming pool, lay the 'Tejon Ranch', Northrop's RCS range, a remote facility buried in a gulley on the southern end of the Tehachapi Mountains. People who found themselves in the vicinity of this place in the dead of night had reported seeing orbs of light within the confines of the range and, inevitably, stories that Northrop was testing recovered alien spacecraft began to circulate amongst UFO buffs. Based on Northrop's highly classified work applying electrostatics to the B-2, these pieces of eyewitness testimony now perhaps had a more plausible explanation.

UFO stories had grown up around the two other radar test ranges in the area – Lockheed Martin's and McDonnell Douglas' (now Boeing's) – a score of miles to the east. Closer still, was Plant 42, the site on the airfield at Palmdale where the Skunk Works' huge manufacturing facility was located. It was there that I was due to meet with and interview Jack Gordon, bright and early after breakfast.

It was sometime in the dead hours that I heard the sound of scratching coming from the lock. I turned towards the handle, no more than four feet from the head of the bed, and saw it move, fractionally, but perceptibly, in the thin sliver of light spilling from a crack in the bathroom door. In my head, I knew exactly what was happening even before the action started. Someone was about to hard-key the front door lock and there was nothing I could do about it. As I braced for the bang from the hammer that would drive the platinum tool into the barrel of the lock and core it out in a split-instant, it happened, but so fast, all I could do was lie there, paralysed, as the door flew in and the bolt-cutters took down the chain. Then, I saw them: three shapes silhouetted momentarily in the doorway – black jump suits, body-armour, kevlar helmets and semi-automatic weapons. A glint of reflected moonlight on the barrel of the gun as it moved and then it was hard against my forehead, grinding against the thin skin between my eyebrows. More pressure, killing pressure, as the snub-barrelled M16 was pushed down harder and harder, the force driving my head deep into the mattress. And above it all, the double-click of the bolt being cocked and locked, a shock of white teeth in a face otherwise hidden by the shadows as he braced to take the shot.

It was then that I came up for air, sucking it down in great gulps, my hands outstretched, pushing against bodies I could still see but weren't there and never had been.

Slowly, the ropey, 70s charm of my motel room at the Desert Inn pulled into focus. I was bathed in sweat and my hands were shaking. I lay there motionless until the sweat dried on my skin and then I tried to get back to sleep, but against the rattle of the air-conditioning unit, I knew I was fighting a losing battle. Around four-thirty, I gave up, got out of bed and stepped outside, a bottle of Coke in hand, into the enveloping warmth of the tinder-dry Mojave night air.

The liquid shimmer of the night sky had been diluted by the first traces of dawn light, dimming the intensity of the stars a little, but not the overall effect. I lay back on the hood of my car and tried to lose myself in thoughts of the cosmos and Garry Lyles' dreams of a '50-year fast trip' to Alpha Centauri in a craft whose modus operandi was beyond the comprehension of the best brain-talent in NASA.

After six months of sniffing around the black world, even longer reporting on its periphery, I had found nothing within the Air Force or its attendant industry to suggest the remotest link with any of the five methodologies espoused by NASA for cracking the gravity code. All I had found was a possible link to T. T. Brown and, significant though this was, it left too many gaps in the overall picture.

While Brown's early work might conceivably have explained how an aircraft with the performance characteristics of a UFO had come to be 'within the present US knowledge' in 1947 – the footnote to the memo I'd found between USAAF generals Twining and and Schulgen (now stained with the coffee I'd spilled across it) – it begged other questions: why NASA had studiously chosen to ignore his work, for example, as a pathway to anti-gravity.

The official reason, according to Valone, LaViolette and others, was that the Air Force had carried out its own investigation into the Brown effect as recently as 1990 and had found no cause to believe it was real. In a report on 'Twenty First Century Propulsion Concepts', prepared for the Air Force Systems Command Propulsion Directorate, the author, Robert Talley, had concluded from his experiments that 'no detectable propulsive force was electrostatically induced'. This failure to replicate Brown's work was cited by both NASA officials and Dr Hal Puthoff as a valid reason for excluding the Brown effect in their search for exotic propulsion technologies.

But what the Air Force report did not make clear was that Brown had consistently advocated using very high voltage to induce an electrogravitic effect – 50 kilovolts in the case of the 2 ft diameter discs and 150 kV in the follow-up experiment that had supposedly exhibited results 'so impressive as to be highly classified.' In Winterhaven, Brown had urged further tests using a 10 ft diameter disc charged at 500 kV – the Air Force, by contrast, had not raised the voltage in its tests above 19 kV.

During two out of dozens of readings, it conceded that it had recorded some 'anomalous' results, but it then made no attempt to explain these anomalies in any subsequent analysis.

I tried to figure why this should be.

Was the Air Force afraid of the truth – or merely trying to suppress it?

In the clarity of thought that can come in the aftermath of dreams, the other notion I latched onto was this: ever since I'd stepped into the black world I'd not been moving forward, as I'd hoped, but backwards; to a period of research I'd already covered, rooted in work that had gone on a half century earlier.

I was trying to find answers in high-tech hotspots like the Skunk Works, when something told me I was looking in the wrong place; the wrong time, even.

And Marckus, I felt sure of it, had known about this from the beginning. He'd been in possession of the facts surrounding the B–2 and for all I knew he was bang up to date on NASA's advanced propulsion activities as well. For the past God knows how long all I'd done was duplicate his vast knowledge in a couple of key areas.

I'd been set some kind of practical exam and gone along with it.

After breakfast, I set off for my appointment with Gordon, keeping the railway on my left and the sun behind me. In the distance, through the early morning haze, a giant hangar pulled into view. It had been built in the 70s as a final assembly hall for Lockheed L1011 airliners, but subsumed since the early 90s by the Skunk Works for the only official work on its stocks: a bunch of 'onesie-twosie' upgrade programmes associated with the U-2 and F-117A.

The hangar dominated the desert scrubland for miles.

I pulled into the car park across the road from the main facility and was met by Ron Lindeke, a seasoned veteran of Lockheed's PR machine. Since the interview was to take place on the main site itself – a rare sanction, Lindeke reminded me – I was reminded to leave my camera and tape-recorder in the PR building. This functioned as an air-lock between the outside world and the organisation's highly secure operations on the other side of the road.

We drove in his car to the main gate, where our documentation was double-checked and triple-checked. It took an age before we were finally admitted to aerospace's inner sanctum.

Gordon's office was located in an administration block in the lee of the giant hangar. There was a low-slung building to the left of the gate and a cluster of busy-looking facilities to the right. These were the factories where the bulk of the company's classified work went on.

The giant hangar was what you were meant to see. The real work went on in the buildings in its immense shadow.

I followed Lindeke to Gordon's office. The cool, low-lit atmosphere chilled the sweat on my back and I wondered what it must be like to work in this windowless environment all the time, day-in, day-out; and whether it was true, as the engineer had told Bill Scott, that they really didn't let you go once you were inducted.

We stepped into an office, as dimly lit as the corridor I had just left. A tall, powerfully built man got up from a table on the edge of the room. Jack Gordon, the fourth president of the Skunk Works, smiled hesitantly and shook my hand.

The Skunk Works and its dynastic leadership system was founded in 1943 by Kelly Johnson, a hugely talented, opinionated and irascible aerospace engineer, in response to an urgent US Army Air Forces requirement for a jet fighter to counter the emergent threat of the Messerschmitt 262. The Me 262, a twin-jet swept-wing fighter-bomber of revolutionary performance, had recently been identified by Allied intelligence as being under development for the Luftwaffe.

Johnson had long been badgering the Lockheed leadership for a special engineering department – small, agile and free of the overweening overheads and administrative baggage that encumbered technical progress in every large aerospace corporation (including Lockheed); and with the XP-80 project he finally got it. The XP-80 was built in just 143 days and, despite arriving too late for a showdown against the 262, went on to become an aviation classic.

The unorthodox methods Johnson pioneered to bring the XP-80 in on time and under budget – all the while under conditions of intense secrecy – set the standard for everything that followed. Ten years later, when the CIA drafted a requirement for a spyplane that could overfly the Soviet Union, it was Lockheed that came up with the solution. When the Agency needed an aircraft to surpass the performance of that aircraft, the glider-like U-2, again it was Lockheed that furnished the answer, this time with the extraordinarily graceful Mach 3 A-12 Blackbird.

Like the U-2, the A-12 started as a black programme, but gradually moved into the light, emerging finally as the USAF SR-71, the fastest operational aircraft in the world until its retirement in 1990. To this day, there is nothing faster, but that, of course, presupposes the 'Aurora' stories aren't true.

Ben Rich, who succeeded Johnson in 1975, denied in his book *Skunk Works*, which serves as its semi-official history, that Aurora ever existed. But Rich did not offer any plausible explanation for what his company had been up to since it produced the last F-117A Stealth Fighter in 1990. There had been roll-outs of two YF-22 prototypes – the forerunner of the USAF's new F-22A Raptor advanced tactical fighter – and a couple of unmanned DarkStar reconnaissance vehicles, but neither of these could account for the comings and goings of 4,000 workers or the busy appearance of the low buildings I had passed on my way in.

I was questioning Gordon about the Skunk Works' operating charter when he let slip that in the course of his 23 years with the company he had worked on 15 'real flying aircraft' – adding hastily that he could only talk about 12 of them.

It was in the ensuing seconds that the thought formed and rapidly solidified in my sleep-starved brain.

Whatever was going on here, however many secret programmes the Skunk Works had on its books, none of them had anything to do with anti-gravity.

Perhaps Aurora was real. Maybe it was the 'fast mover' that people had described crossing the high frontier above the South-Western US as a pulsating point of light – as reported in *Aviation Week* in the late 1980s. Maybe it employed a drag-reduction device of even greater sophistication than the one apparently on the B-2. But whatever it was or wasn't, Aurora was an aircraft, designed and developed by aerospace engineers.

Forty years earlier, T. T. Brown had proposed specifically that Project Winterhaven should be built outside the aerospace industry. Maybe Brown knew intuitively what it had taken me months to grasp. The aerospace industry had more to lose from anti-gravity than it could ever gain from it. Gordon's entire business base relied on the science of aerodynamics. Take that away and he, it – everything about the aerospace industry, in fact – ceased to have any role or purpose.

Anti-gravity threatened to wipe the aerospace industry off the map.

Was this the reason for the crushing silence that had followed the industry's initial burst of enthusiasm for anti-gravity in 1956? If so, it was the simplest explanation of all. And I had spent months on the wrong heading.

I pressed Gordon on the missing three aircraft, but in view of the direction my thoughts had taken, I must have unintentionally signalled my indifference.

When he raised his eyes to mine, they told me that the meeting was over.

Minutes later, Lindeke escorted me back towards the high heat that was baking the floor of the Antelope Valley.

Just before I left the building, I stopped in front of a large chart on the wall of the lobby area. I hadn't noticed it on the way in.

It proudly illustrated the lineage of every Skunk Works aircraft since the XP-80. Past the picture of the U-2, past the SR-71 Blackbird and the F-117A Stealth Fighter, past the YF-22 and DarkStar, and there was something called 'Astra'.

Sitting at the top of the tree, Astra looked like an ultra-high-speed reconnaissance aircraft; every pundit's dream of how Aurora ought to look.

It was difficult to gauge Lindeke's reaction because he did not remove his sunglasses, but he knew as well as I that the Skunk Works had built nothing since DarkStar; and DarkStar, being unmanned, looked nothing like 'Astra'.

In the car, I pressed him for an answer and he promised to come back to me. It was several weeks later, long after I returned to England, that he finally did.

'Astra was a high-speed airliner we studied in the 1970s that got canned,' he announced without any preamble. 'I checked.'

It took me a moment to absorb this. 'The thing I saw was at the very top of the chart, Ron, and it was most certainly not an airliner.'

'Hey,' he interjected testily, 'the only Astra anyone here ever heard of was something we considered more than 20 years ago and never built. Take it or leave it.' For a moment, the receiver filled with the sound of his breathing. 'Call me if you got any further questions, won't you?' And with that he was gone.

Chapter 14

Two final way-points on the invisible network of nodes and links that held the black world together and here was the first of them, the Pentagon, with its long, wide corridors leading to the offices of Lieutenant General George Muellner, principal deputy, Office of the Assistant Secretary of the Air Force for Acquisition. After a month's interlude in the UK, I was back in Washington on official business.

It was Muellner's last day as the service's chief procurement officer. Tomorrow he retired, the culmination of a service career that had begun in the late 60s as a pilot flying F-4s out of Cam Ranh Bay, South Vietnam.

According to his bio, he had more than 5,300 hours on F-4s, A-7s, F-15s and F-16s. Muellner completed 690 combat missions over Vietnam and logged another 50 combat sorties during the Gulf War in 1991, at which time he commanded a deployment of Joint STARS battlefield surveillance aircraft. In 1991, Joint STARS was still officially under experimental test, but its belly-mounted radar proved invaluable in picking out Iraqi tanks and Scud launchers against the flat backdrop of the desert. It was, to use the jargon, an 'enabling technology' that had helped the Coalition to win the war.

Another 'enabler' had been stealth, a technology with which Muellner was intimately familiar, although you would never guess it from his official past. The merest clue that he had once been bound up in the stealth revolution came in the fourth paragraph of the 'assignments' section of his bio, the part that outlined the key milestones of his career. While the maximum period of time contained in each of the 17 other paragraphs covered a year, two at the most, the fourth paragraph spanned an unusually long length of time: May 1973 to July 1982.

During these nine years, Muellner's activities were described as 'F-15 operational test pilot, 4486th Test Squadron, Edwards Air Force Base, Calif; chief of systems testing, USAF Test Pilot School; test pilot with the F-16 Combined Test Force; and operations officer and commander, 6513th Test Squadron, testing classified aircraft.' It was kind of like Pittman Station all over again.

So, Muellner had the Right Stuff in spades and he had lived a great deal of his professional life in the black. His nine years in the test community, it turned out, covered one of the most intense periods of development in the US Air Force's history – a time when it had had to take a long hard look at its equipment and tactics and rapidly come up with a raft of new ideas and hardware to ensure that it was as ready for the next conflict as it had been ill-prepared for the one it had just fought and lost in Vietnam. It was this period that presaged the biggest hike in black world spending since the days of the first and most grandiose black world project of all: the bomb.

By even stepping into the building – the great edifice constructed by General Leslie Groves prior to his appointment as director of the atomic bomb project – I knew I had entered a world of Catch-22s. If the Air Force was developing anti-gravity, there was no way, of course, that Muellner was simply going to come clean and admit it. Then again, if it was going on, it was also possible Muellner had been excluded from it. If this was the case, Muellner's answers to my questions, whether they were true or not, would have 'plausible deniability' – the earnest and sincere response of an official whose résumé said he should be in the know, even though in reality he wasn't.

It seemed an apposite way to start, so I asked the general whether he was satisfied that he'd been briefed on everything that was being procured for the Air Force in his name.

He replied without missing a beat. 'I believe so. You know, it's one of those questions you never know the answer for sure, but I think the answer is "yes".'

So, just what was the Air Force developing in the black?

'Things that push the envelope beyond where we're at,' he replied, again without hesitation. Next-generation stealth, certain breakthrough weapons technologies, information warfare, spy satellites. That sort of thing.

Civilians who made it their business to keep tabs on such affairs, sleuths like the Desert Rats who ventured as deep as it was possible go into black world territory without getting their heads blown off, claimed that there might be as many as four secret 'platforms' flying out there, I probed, including the fabled Aurora – or was it Astra now?

Suddenly, Muellner looked uncomfortable in his stiff, blue uniform; a shirt-sleeves airman who'd have gladly swapped the formality of his Pentagon suite for something a little closer to the front line of advanced flight test.

'If they are,' he said, 'then my earlier answer about knowing about all our programmes may be faulty, because I'm certainly not aware of them.'

The Air Force, he added, had no need for a reconnaissance aircraft that could fly around the globe at Mach 7 or Mach 8 when the satellites it already had in place could, for the most part, do the same job at a fraction of the cost.

Stealth, however, was a different matter and here the Air Force was developing a range of new technologies in the black.

Muellner wouldn't be specific, but he hinted that at least part of this effort entailed the development of 'reactive skins' that would enable a next-generation stealth fighter to change colour like a chameleon, blending against its background from any perspective. A cloaking device that could, in effect, make an aircraft disappear into the surrounding ether. Visual stealth to add to the radar and infrared variety already in place.

None of these efforts had resulted in any new operational aircraft yet, but there were test programmes under way, he said, designed to 'mature technology'.

From this it was acceptable to deduce that *something* had been developed at the time of the Aurora sightings over the North Sea oil rig and elsewhere, that something real had been tripping the seismology sensors of the USGS, but that for whatever reason the vehicle had not been procured for front-line service. Why? Because Muellner didn't need to be saying any of this. The Air Force never talked about black programmes; but on this, his last full day in office, had the general decided to invoke his other responsibility as the service's 'chief information officer' and set the record straight?

The black world had taken a pasting of late. The rumours about what the Air Force was doing away from the public light of accountability were starting to spiral out of control. There were certain things that the Air Force would always need to develop in the black to maintain America's readiness in an uncertain world, Muellner said.

'And the rest?' I asked him, easing between the agreed subject-matter of the interview and the anti-gravity question. What about the people – apparently sane and credible witnesses – who said that they saw things in the desert night skies that seemed to defy the laws of physics? Craft that hovered right over them sometimes, huge, motionless, silent?

Muellner smiled. Get him the name of the contractor, he said, because he'd love to put them under contract. Whatever was out there, it had nothing to do with the Air Force.

But here was the disconnect. People *were* seeing things that defied explanation in the vicinity of Area 51, Nevada, the US Air Force base that didn't exist a hundred miles north of Las Vegas, north of a bend on Highway 375 – the 'Extraterrestrial Highway', as it had been dubbed by the Nevada state legislature.

Here, Tikaboo Valley, 25 miles wide and six times as long, as lonely a place as you could find on the planet, stretched into the distance till it merged with the horizon.

It was Tikaboo Valley that connected the outside world to Area 51, hidden as it was by the jagged peaks and ridges of the Groom Range and the Jumbled Hills. Right above it, high on top of Bald Mountain, elevation 9,380 feet, sat the radar site and observation post that gave security officers an unrestricted view of all traffic entering and transiting the area. It was the first of several security rings and sensor networks employed by the Air Force to ensure that nothing penetrated the discreet *cordon sanitaire* around the base. The others, if you were unlucky, could be a lot less passive, more in-the-face: the sworn-in deputies from 'Pittman Station' who patrolled the valley looking for miscreants and who occasionally locked them up and were said to throw away the key; and the Black Hawk helicopters that had been known to sand-blast the odd Desert Rat from his hidey-hole on the high ground overlooking Area 51's 27,000 ft runway.

Already one of the longest paved runways in the world, it had been specially extended in the late 1980s in a move that supported the widely held view in the 'stealth-watcher' community that a very high-speed aircraft was under test at Area 51. Couple this with the unmarked charter planes and buses with their windows blacked out that were ferrying up to 4,000 workers in and out of the base from Los Angeles and Las Vegas every week and word soon got around that Area 51 was as busy as it had ever been in its 40-year history – ever since, in fact, Kelly Johnson, the Skunk Works' founder, brought the U-2 here for testing in the mid-1950s.

But alongside the stealth-watchers, another set of pilgrims regularly made the trip to Tikaboo Valley. Many of these people had been drawn to Area 51 by the revelations of a man named Robert Lazar, who in 1989 claimed to have worked on recovered alien spacecraft at a site adjacent to Area 51 known as S-4. While Lazar's claims seemed wild, his followers consistently reported sightings of strange orbs of light over the base; objects that seemed to defy the way aircraft – even the top-secret variety at Groom Lake – ought to fly.

But tread the fine line between the world of black programmes and the world of UFOs and you entered another minefield of disinformation.

In 1997, the CIA admitted in a set of released documents that it had encouraged reports of flying saucers in the 1950s and 1960s to obscure the flight-trials of its then top-secret U-2 and A-12 spyplanes. The implication was that if you saw one of these things overflying the highway and reported it, there was something wrong with your reasoning. It was a tactic that worked well.

Add this to the highly compartmentalised, watertight environment in which black programme engineers and scientists worked, and the fear they were reportedly subjected to, and here, as with the Philadelphia Experiment, was a cheap, but highly effective means of maintaining programme secrecy. By the cusp of the millennium, there was so much bad information mixed in with the good about black programmes that it was often impossible to tell them apart.

Was Astra another case in point?

On my return from California I'd left a message on Marckus' answer-machine. Maybe, I'd conceded, talking to the machine like it was a confessional, I had been expressly meant to see that chart. Maybe Astra was a subtle piece of disinformation; a misleading fragment planted ahead of my visit. Then again, maybe it wasn't. Maybe it was genuine.

Faced as I was with the official denial, Astra was every bit as effective as a hit team equipped with Colt Commandos and a platinum hard-key. Less messy, too.

This time, Marckus had called me back. I told him then what I thought of his wild goose-chase. He never attempted to deny it. I got the impression he truly believed it was good for me, like it was an extension of my education or something.

In the small hours of a particularly hot night, three days after I'd sat in Muellner's office, I watched from my mountain-top look-out as a ball of shining golden light rose above the hills that separated Groom Lake from Tikaboo Valley. The light appeared to hang in the air for a few moments before drifting lazily downrange and disappearing behind a mountain peak 50 miles to the south.

It left me strangely unmoved.

In the kind of silence that you can shred with a razor when you're a hundred miles from anywhere, I found myself thinking again about the memo from Lt Gen Nathan Twining, head of the US Army Air Forces' Air Material Command, to Brig Gen George Schulgen, chief of USAAF intelligence in September 1947. The memo I'd spilled coffee over in my basement back in London.

Twining's observation – that objects 'approximating the shape of a disc, of such appreciable size as to appear to be as large as a man-made aircraft' were neither 'visionary nor fictitious' – was written just three months after the supposed crash of a UFO at Roswell, New Mexico, in July 1947: the period that most people tout as the real start of the modern UFO sightings era.

What was it that Twining had written? That it was possible, 'within the present US knowledge – provided extensive detailed development is undertaken – to construct a piloted aircraft which has the general description of the object above which would be capable of an approximate range of 7000 miles at subsonic speeds.'

In mid-August 1945, I recalled, too, that 300 railroad freight cars of V-2 components captured in the European theatre of operations had trundled westwards across the wilderness from the Gulf of Mexico to the Organ Mountains, deep inside the dry, desert hinterland. Every railroad siding from El Paso, Texas, to Belen, New Mexico, a distance of 210 miles, had been seen to be full of cars all headed for the White Sands Proving Ground, 125 miles to the west of Roswell. Following on behind them were approximately 40 captured German scientists recruited under the covert selection programme known as 'Operation Paperclip'. They were led by Wernher von Braun.

Maybe Roswell had happened, maybe it hadn't. But if something had come down in that summer of 1947, something extraterrestrial, there was no definitive proof, more than half a century later, to say that what had crashed there was alien; and dozens, if not hundreds of researchers had sought to discover the truth about it.

But if Roswell's extraterrestrial dimension was simply a piece of fiction, then why not turn the thing around, look a little to the left or the right, and see what swam into focus.

When I looked at Roswell this way, I saw the White Sands Proving Ground. And then I saw something else.

The V-2 rocket programme had started under the aegis of the German Army, the Wehrmacht, but had been transferred in the final stages of the war on Hitler's orders to a higher authority: Reichsführer-SS, Heinrich Himmler. The same had been true of the Me 262 jet fighter and just about every other German high-tech weapons programme. Most of them had started under Luftwaffe control and then migrated in the same direction. In short, anything that had shown any real promise as a weapon system – in particular, anything that appeared to represent a quantum leap over the then-state-of-the-art – had ended up under the oversight of the SS.

Eight thousand feet above sea-level, I felt a constriction in my throat. I was so keyed, my breath was coming in short, sharp gasps. But the thoughts kept on tumbling.

Just as it seemed I'd been wrong in trying to connect anti-gravity with the USAF, so I realised, in a moment of clarity, I'd made the same mistake in Germany.

What had *Lusty* actually stood for? *Luftwaffe* Secret Technology.

In looking for evidence of an anti-gravity programme in the raw intelligence reports of Project *Lusty*, had I been heading up a blind alley?

If anti-gravity had been developed by the Nazis, then perhaps it hadn't been under Luftwaffe control at all, but that of the SS.

I called Marckus from a call-box in a windswept diner just after I turned off the Extraterrestrial Highway for Vegas. I told him the conclusions I'd reached and that the available clues said I should take another look at Germany.

'I think you're making a mistake,' Marckus said. 'If anyone has cracked it, it's the Americans. Stay in the States, use your contacts, go deeper.'

I neglected to tell him that, thanks to his field test, I'd wasted enough time already.

Chapter 15

In the reading room of London's Imperial War Museum, under a skylight battered by autumn winds and icy rain, I immersed myself in books and papers that plotted the SS' gradual take-over of the German aerospace and weapons industry in the last two years of the war. But as the evidence mounted, I found myself increasingly distracted by thoughts of Marckus. Why had he warned me off the German research angle? Why did Marckus, with his incisive mind and razor-sharp reasoning, insist on belittling information that seemed more beguiling by the hour?

It was the intractible aspect of this contradiction that led me to pledge quietly to myself that from now on I'd be much more careful about what I told him.

It was in August 1943, just three months after the first successful test shot of the A-4 rocket, better known as the V-2 vengeance weapon, that Himmler, head of the SS, managed to wrest the V-2 from the control of the Wehrmacht, the German Army. Himmler was at the Wolf's Lair, Hitler's remote headquarters in the forests of East Prussia, when the Führer, in the company of his deputy Martin Bormann, brought up the subject of A-4 manufacture and his concerns about programme secrecy. Hitler, as was his wont, was demanding the impossible.

The recent test of the A-4, which had landed just five kilometres off-target over a distance of 265 kilometres, had finally convinced Hitler, after months of vacillation, that the rocket would thwart the Allies' developing invasion plans and quite possibly bring Britain to its knees. He demanded that 5,000 rockets should go into immediate production and be delivered in the shortest possible time.

Himmler saw his chance. Quick to exploit an opportunity, the Reichsführer knew that neither the Wehrmacht nor Albert Speer, Hitler's young and energetic minister of armaments and war production, could deliver the project under existing arrangements. Hitler had decreed that 'only Germans' should be employed in the production of the A-4, but the sheer size of the task meant, practically speaking, that foreign civilian workers – Poles, Hungarians, Czechs and others from the conquered east – would have to be involved sooner rather than later in the process. But with foreign workers came the risk of the security leaks that Hitler had expressly sought to eliminate. It was to this apparently insoluble conundrum that Himmler claimed he had the answer.

What he offered Hitler, he promised, was watertight: his proposed workforce would be drawn not from the diminishing manpower resources of the Reich's civilian labour-pool, but from the SS-controlled concentration camps, from which there was no escape. The same work-force, Himmler added, could be used to build the underground manufacturing facilities that would keep the A-4 secure from Allied bombing. Faced with the logic of this proposal, Hitler agreed, but informed Himmler that he must cooperate with Speer.

Though the SS controlled the camps, it was the armaments ministry that had masterminded the revitalisation of the German manufacturing economy, achieving month on month growth, in spite of the ferocity of the Allies' strategic bombing campaign, since Speer's appointment the previous year. Yet, when he was informed of Himmler's proposal, Speer couldn't help but object to it, knowing that Himmler, whose dreams invariably exceeded his abilities, would never be able to deliver what he had promised. This was undoubtedly the case, but it was not what Hitler wanted to hear. The following day, Himmler informed Speer that he had taken charge of the manufacture of the 'A-4 instrument' (as it was known to conceal its identity and purpose to outsiders) and demanded Speer's full cooperation as a subordinate. This the outmanoeuvred armaments minister was forced to provide.

Thus, within 24 hours, did Himmler not only acquire control of the V-2 programme, but in the process lay the foundations for the SS take-over of all secret weapons programmes within the Third Reich.

For years, the Reichsführer-SS had harboured a grand scheme to turn the SS, originally established as Hitler's bodyguard, into a self-sustaining economic entity, capable of funding and running its affairs without reliance on the everyday financial and administrative mechanisms that maintained and sustained the Reich. It was via the camps that Himmler and his leading economic authority, Oswald Pohl, realised that they could achieve their aims. By 1943, population of the camps had surged – from around 25,000 on the eve of the war, to well over half a million, a figure that did not begin to reflect the millions that had already died in death camps like Auschwitz, Dachau and Mauthausen and the millions more that would die of disease, starvation and the gas-chambers before hostilities were over.

In a perverse reflection of the growth of the camps, the size of the SS itself surged from an eve-of-war strength of 240,000 to an end-of-war tally numbering 800,000. Aside from its remit to maintain security within Hitler's state and its brutal administration of the death and labour camps, the SS also provided 38 fighting divisions alongside the Wehrmacht.

After the abortive bomb-plot against Hitler in July 1944, a scheme hatched by anti-Hitler elements within the German Army General Staff, the control of the SS and its grip upon Germany would become absolute. But it was in August 1943, at the Wolf's Lair, that the seeds of the SS' take-over of the armaments industry were sown. Skilled workers and scientific specialists, Himmler told Speer, would immediately be transferred from the camps to begin work on the A-4. Furthermore, the Reichsführer-SS added, he had asked a 'young energetic construction expert, who had already proved his outstanding ability, to take charge of the enterprise.'

Speer must have had an overwhelming sense of inevitability as to the identity of this individual, but he inquired anyway. The man who was to assume direct oversight of the A-4's manufacture, Himmler told him, was Dr Ing (Engineer) Hans Kammler, head of the SS' Building and Works Division – the entity that had masterminded and built the camps.

Kammler's rise inside the SS had been meteoric. Up until 1941, he had been an unremarkable civil servant within the RLM, the Reich Air Ministry, with special responsibility for construction projects – hangars, barracks, administrative offices and the like. In the autumn of that year, realising that he was never going to rise to the heights to which he aspired, he transferred to the SS, where he received the rank of Brigadeführer and was immediately put in charge of building projects. By the end of that year, Kammler had already drafted a five-year RM 13 billion programme of construction for SS barracks and concentration camps stretching from the newly captured territories of the Soviet Union to Norway.

The sense that the SS gave Kammler free rein to indulge his ideas and ambitions – suppressed during his years of obscure service at the RLM – is unavoidable. No sooner had his appointment taken effect, than he was drawing up detailed plans for the rapid expansion of the camps, egged on by Pohl, who had impressed upon him their value as the engines of growth for his and Himmler's grotesque vision.

To turn it into reality, Kammler immediately saw what needed to be done and requested to expand the capacity of the camps to four million; three months later, he increased this figure again to 14 million.

While the building programme got under way, he set about industrialising the existing camps, but progressively found himself thwarted by

Speer, who, irrespective of any feelings he may or may not have had about the moral consequences of such a scheme, believed that the camps were not the answer to Germany's manufacturing needs. Yet it was a confrontation that Speer was bound to lose, as he tacitly admitted in his memoirs; for, in Kammler, he was essentially battling against a twisted image of himself.

'He too came from a solid middle-class family, had gone through university, had been "discovered" because of his work in construction, and had gone far and fast in fields for which he had not been trained.' Speer, it is clear, was jealous of Kammler, but while Speer patently wrestled with his conscience, especially during these years, there is no sign of any moral doubt on the part of the 40-year-old SS officer.

Kammler was cut from the same cloth as SS-Obergruppenführer Reinhard Heydrich, the author of the 'Final Solution of the Jewish Problem', who was assassinated by Czech agents in Prague in 1942.

'Both were blond, blue-eyed, long-headed, always neatly dressed, and well bred. Both were capable of unexpected decisions at any moment, and once they had arrived at them would carry them through with a rare obstinacy,' Speer noted.

With Heydrich dead, Himmler had found in Kammler a cold, ruthless and energetic executor of his wishes; the man who would carry though his simple, yet grandiose vision.

Kammler set about his task as overseer of the A-4 manufacture programme with clinical zeal. One week after Himmler's fateful meeting with Hitler at the Wolf's Lair, he dispatched the first group of concentration camp inmates from Buchenwald to Nordhausen in the Harz Mountains of central Germany, where work on a giant underground construction facility for the A-4/V-2 commenced.

Within a month, Kammler and Speer, the latter in his new capacity on the project as junior partner to the SS, established the low-profile Mittelwerk GmbH (Central Works Ltd) to run the rocket venture.

By the end of the year, 10,000 Buchenwald prisoners – mostly Russians, Poles and Frenchmen – had been dispatched into the limestone cliffs of the Kohnstein, the mountainous ridge close to the village of Nordhausen into which the facility was tunnelled, to bring about the impossible: the construction of the largest underground factory in the world, a facility a kilometre and a half long containing 20 kilometres of tunnels and galleries, dedicated to the construction of rockets, missiles and other top-secret weapons. It was finished in a year – 20,000 prisoners dying in the process.

Plans show that it was the first phase in a construction programme that would have made the Mittelwerk part of an enterprise three times the size.

Thereafter, Kammler's ascent to the top levels of the bureaucracy surrounding Hitler's inner cabinet was unstoppable. As 1943 drifted into 1944, the SS' steadfast refusal to believe in anything other than final victory ensured its deepening involvement at all levels of state. In March 1944, at the formation of a combined 'Fighter Aircraft Staff' composed of senior Luftwaffe and armaments ministry officials tasked with increasing fighter production in the face of the Allies' determination to wipe it out with carpet-bombing, Kammler represented the SS as Himmler's special delegate, offering two primary areas of expertise: his rapidly developing knowledge of underground factory construction and his ability to mobilise vast numbers of concentration camp labourers. In recognition of this, Reichsmarschall Hermann Goering, head of the Luftwaffe and Hitler's nominal successor, tasked Kammler with transferring strategic aircraft factories underground. Just over three months later, Himmler was able to report to Hitler that ten aircraft factories with an underground floor space of tens of thousands of square metres had been constructed in just eight weeks.

As the Reich's decay accelerated under the relentless news of defeat on all fronts – the D-Day invasion had placed Allied troops in France, the Red Army had just crossed into Poland and British and American bombers were pummelling German cities daily from the air – Kammler's influence spread. In the air of paranoia that followed the bomb plot of 20 July, his ascendancy entered its final and most decisive phase. On 6 August, following Himmler's appointment as head of armaments, adding to his portfolio as chief of internal security and commander-in-chief of the reserve army, Kammler was given responsibility for all aspects of V-2 operations – from design through manufacture to deployment and offensive operations against England and the Low Countries. In effect, Kammler now had sole charge of the project from planning to firing; a position that, thanks to his obsessive eye for detail, gave him an insight into the running of a strategic weapons programme that no other individual has enjoyed before or since.

In January 1945, with the Red Army in Budapest and Warsaw and the Western Allies turning the tide of the Germans' Ardennes offensive, Kammler was appointed head of all missile programmes – defensive as well as offensive. On 6th February, Hitler transferred cradle-to-grave responsibility for all air weapons – fighters, missiles and bombers – to the man that many Party insiders now viewed as the most powerful and influential state official outside Hitler's inner cabinet.

Kammler had first come to Hitler's attention with a carefully crafted hand-coloured design for Auschwitz concentration camp. It was his perverse capacity to retain control of the minutiae of a project without ever relinquishing grip on the strategic objectives, a trait shared by Hitler himself, that drew him increasingly into the Führer's orbit; a characteristic that came to the fore when Kammler 're-engineered' the Warsaw Ghetto after the crushing of the uprising in August 1944. He not only levelled the 'contaminated area', but made sure that each of the 34 million bricks from the Ghetto was used for other essential building projects. He was also called in to advise on methodologies for increasing the daily output of the camps' gas-chambers and ovens from 10,000 to 60,000. For this and his other work he was rewarded with an SS general's rank of Obergruppenführer.

By the beginning of April 1945, with the Red Army driving towards Berlin and Hitler now ensconced in the Führerbunker below the Party Chancellery, Kammler had acquired control of every single major aircraft – or rather, *aerospace* – programme of any importance within the Third Reich. His empire now included any project or facility that had either an SS influence or was aircraft or missile related. Geographically, this gave him the authority to enter and operate any weapons plant within the shrinking borders of the Reich, by now restricted to Norway, Denmark, northern Germany, southern Bavaria, Czechoslovakia, Austria and northern Italy.

What is astounding is the belief of the Nazi leadership, even at this terminal stage of the war, that Kammler would bring about miracles and deliver the German people from the relentless advances of the Soviets in the east and the British and Americans in the west. 'The Führer has had prolonged discussions with Obergruppenführer Kammler who now carries responsibility for the reform of the Luftwaffe,' propaganda minister Joseph Goebbels wrote in his diary on 3 April. 'Kammler is doing excellently and great hopes are placed on him.'

Yet, for all the power vested in Kammler at this late stage of the war he barely rates a mention in standard works of reference on the Luftwaffe or its major programmes. There is so little data on him, in fact, that archivists at London's Imperial War Museum confessed that their files contained next to nothing on him.

Just when I thought I was going to have to look elsewhere, perhaps in Germany itself, I received a phone call from a diligent researcher at the IWM who'd become as engrossed as I was by the lack of information on Kammler.

The researcher told me, a note of triumph in his voice, that he *had* turned up a book on Kammler after a rigorous search: a slim volume, with a curious title – *Blunder! How The US Gave Away Nazi Supersecrets to Russia* – penned by a British Cambridge graduate-turned-foreign correspondent called Tom Agoston.

What *Blunder!* contained, however, was dynamite, for Agoston had established what no other researcher has managed to before or since.

While Kammler carried out his remit to the letter, churning out the rockets and jet aircraft that Hitler hoped would turn the tide against the Allies in the closing weeks of the war, he also set up, unbeknownst to anyone connected with those projects, a top-secret research centre tasked with the development of follow-on technology, a place where work on 'second-generation' secret weapons was already well-advanced.

'In modern high-tech jargon, the operation would probably be referred to as an "SS research think-tank",' Agoston wrote.

In the jargon of the US aerospace and defence industry, however, Kammler's 'think-tank' would have been known by a different term – one that would have been familiar, though, to anyone who had ever been consigned to the black world.

What Kammler had established was a 'special projects office', a forerunner of the entity that had been run by the bright young colonels of the USAF's stealth programme in the 1970s and 1980s; a place of vision, where imagination could run free, unfettered by the restraints of accountability. Exactly the kind of place, in fact, you'd expect to find anti-gravity technology, if such an impossible thing existed.

Chapter 16

According to Agoston, Kammler set up his special *Stab* (staff) for these visionary projects in a highly compartmentalised section of the Skoda Works in the Reich Protectorate of Bohemia and Moravia in the vassal Nazi state of Czechoslovakia. Through the brutal administration of Heydrich, the Protector of Bohemia-Moravia until his assassination in June 1942, the SS had come to treat the industry of the entire Czech region as its own private domain. In March 1942, Himmler formally took over the operations of the Skoda Works, a giant industrial complex centred on the Czech towns of Pilsen and Brno, as an SS-run operation. In the process, he managed to bypass Speer, who knew nothing of the arrangement until Hitler told him it was a *fait accompli*. With the SS in charge of Skoda, Hitler tasked Himmler with evaluation and, where appropriate, duplication of captured enemy weapon systems whose performance outstripped the Germans' own. Under the leadership of Skoda's director general, honorary SS-Standartenführer (colonel) Wilhelm Voss, this project was completed in record time. On 11 May 1942, Voss was able to explain to Himmler what it was that had enabled him to progress the development of these back-engineered weapon systems so successfully – 'by concentrating all developments for the Waffen-SS in the liaison staff of Skoda and Bruenn [Brno] weapons plants, and assuring systematic, intense cooperation with the SS Ordnance Office.'

In effect, Voss was telling Himmler that he had been able to move as quickly and effectively as he had thanks to the removal of the overweening bureaucracy that went hand-in-hand with projects administered by Speer's armaments office.

As far as the regular mechanisms of the Reich procurement machine were concerned, then, the Skoda operation was to all intents and purposes beyond oversight. This would set the scene for the Kammler special projects group, which was so deeply buried, so beyond regular control, it was as if it did not exist.

Voss, 'tall, quietly reserved in manner, greying, upright and elegant', would become Kammler's alter-ego in the administration of the special projects group. Seven years older than the SS General and with a reputation as an industrialist that was as solid as a rock, Voss, with his nuts-and-bolts background, was the ideal complement to Kammler's driven bureaucrat. Even better, the two men had a remit from Hitler and Himmler to run the special projects group without the least interference from Goering or Speer. Apparently, neither Goering, head of the Luftwaffe, nor Speer, head of all armaments programmes, would have the least idea it existed.

But for Tom Agoston, in fact, the reality of the *Kammlerstab* might never have emerged at all.

Agoston was a Cambridge graduate who had served during the war as an air photo interpreter. After the war, he travelled to Germany to report on the Nuremberg war crimes trials. It was there that he ran into Voss.

Though not indicted by the war crimes tribunal, Voss was so close to the Nazi elite that his name was constantly on the lips of the prosecutors.

'Everyone wanted to interview Voss, but they couldn't find him,' Agoston told me over the phone from his home in Germany a half century later, the satisfaction still evident in his voice. 'As he was sheltering in my house at the time, this was hardly surprising.'

In the course of several extended interviews in 1949, Voss unburdened himself of the whole Skoda story, speaking of his activities there with 'unique frankness'.

Tired and disillusioned, Voss was patently grateful for Agoston's company and when the subject of the Kammler Group surfaced, he told the reporter everything.

The Kammler special projects office was regarded by the select few who knew about it as the most advanced high-technology research and development centre within the Third Reich. It was totally independent of the Skoda Works' own R&D division, but used Skoda for cover.

All funding for the programmes at the *Kammlerstab* was channelled through Voss, who reported alongside Kammler directly to Himmler. The scientists were culled from research institutes all over the Reich, chosen for their acumen as engineers and scientists, not for their allegiance to the party.

'Many scientists, anxious to see their work in print, even if it was kept top secret, prepared papers for a central office of scientific reports, which circulated them to specific recipients. Some of these reports were used as the basis for selecting candidates for employment at Skoda,' Agoston wrote.

Once recruited to the special projects group, whether they liked it or not, the scientists set about their work, their activities protected by a triple-ring of security provided by SS counter-intelligence specialists assigned specially for the task.

These security rings were established by Himmler around Skoda's sites at Pilsen and Brno and their administrative centre in Prague.

Voss described the activities of the scientists at the *Kammlerstab* as beyond any technology that had appeared by the end of the war – working on weapon systems that made the V-1 and the V-2 look pedestrian. Amongst these were nuclear powerplants for rockets and aircraft, highly advanced guided weapons and anti-aircraft lasers. The latter were so far ahead of their time that by the end of the 20th century there was still no official confirmation of laser weapons having entered service, despite the best efforts of the Russians and the Americans.

In March 1949, several weeks, after he briefed Agoston, Voss was pulled in for questioning by the US Counter-Intelligence Corps.

Shortly afterwards, he wrote in secret to Agoston and requested that *die heisse angelegenheit* – 'the hot matter' – they had discussed earlier that year was not for publication.

Voss had told the CIC agents about the range of research activities that had been pursued by the special projects group and they informed him, in no uncertain terms, that he was never to speak about the *Kammlerstab* or its programmes to anyone.

Voss gave them his word, neglecting to tell them, of course, that he had already briefed Agoston.

Agoston duly honoured Voss' request to put a lid on the matter, but with Voss' death in 1974 he was absolved of his obligation and began picking up the threads of the story again. Voss' interrogation reports should have been freely available under the US Freedom of Information Act, but despite persistent requests, Agoston was told they were unavailable; that there was no record, in fact, of any such interrogations ever having taken place.

I recalled from the *Lusty* documents I had examined in Washington a report that had been filed by OSS agents, forerunners of the CIA, from Austria in April 1945. Bundled into a collection of unprocessed notes on Nazi high-technology weapons 'targets' examined by US intelligence specialists, the report had mentioned a location in Vienna where experiments had supposedly been conducted on 'anti-aircraft rays' – laser weapons.

'Research activity is conducted in a house at the above address (87, Weimarerstr. Vienna). Research personnel were not allowed to leave house (reported hermetically sealed).'

Funny, I thought, how this much later on in the investigation I saw things through a markedly different prism. At the time, I'd been focused on the technology. Now, the fact that the Germans had been working on directed energy weapons was subordinate to a subtly different observation: that the Nazis had operated a compartmentalised security structure like the one now operated by the black world.

It was then that I saw the Kammler special projects group for what it was; not as a facility where experiments were conducted, but as an R&D coordinating centre. Testing wouldn't have been performed at Skoda itself, but in the field.

It was exactly as Agoston said it was: a think-tank, not an experimental site.

I put out feelers into the Czech Republic and Poland. I now saw that if there had been any truth at all in the reports of flying saucers developed by the Germans at the end of the war, this was the place to start looking. If the legend that had grown up around the supposed activities of Messrs. Schriever, Habermohl, Miethe and Bellonzo was in any shape, way or form based on truth – and while my earlier research had shown there wasn't a shred of solid evidence to say they were, I was convinced there was no smoke without fire – two places always seemed to recur in so-called testimony.

The first was Prague, not a million miles from Skoda's twin hubs at Pilsen and Brno (Bruenn). The second was Breslau in the German province of Lower Silesia. Then part of Germany, Breslau – renamed Wroclaw – is today in south-west Poland.

I turned my attention back to Kammler.

In March 1945, Kammler moved his headquarters from Berlin to Munich, working out of the regional Waffen-SS and Reich Police Construction Office. On 16 April, three weeks after he had received his mission from Hitler to change the course of the war, Kammler delegated Gerhard Degenkolb, industrial plenipotentiary for the manufacture of jet aircraft at Speer's ministry for armaments and war production, to assume special responsibility for manufacture of the Messerschmitt 262, Germany's last-ditch hope in the war in the air. On 17 April, with the Third Reich crashing around his ears, Kammler sent a message to Himmler at SS Command Headquarters in Berlin denying Himmler the use of a heavy truck that the Reichsführer-SS had requested from the Junkers aircraft factory motor pool – an order so obsessive in its attention to detail, given the inferno that was burning around Kammler, that Speer later described it as 'both terrifying and laughable'.

Be that as it may. It was the last official message anyone ever received from Kammler, who by 18 April had dropped off the map – effectively without trace.

'In the course of my enforced collaboration with this man,' wrote Speer of Kammler after the war, 'I discovered him to be a cold, ruthless schemer, a fanatic in pursuit of a goal, and as carefully calculating as he was unscrupulous.'

If there was one thing I had learned about Kammler in the short time I had spent analysing his activities, it was that nothing in the clockwork routine of his life happened by chance. By the beginning of April 1945, whilst Hitler was heaping new responsibilities on his shoulders almost by the day, the general would have been acutely aware that it was just a matter of time before Germany collapsed.

In contrast to Himmler, who vainly believed that he could enter into armistice negotiations with the Allies and emerge from them with a leadership role in the reconstituted Germany, Kammler was a realist. His part in the construction of the concentration camps, the clinical way in which he had sought to boost the throughput of the gas-chambers and ovens, not to mention the methods he had employed in levelling the Warsaw Ghetto, would have placed him high on the list of SS officers sought by the Allies for war crimes.

Unlike Himmler, however, Kammler had something of value to deal – something tangible. By early April, Hitler and Himmler had placed under his direct control every secret weapon system of any consequence within the Third Reich – weapons that had no counterpart in the inventories of the three powers that were now bearing down on central Germany from the east and the west.

A man who had no trouble transforming the grand visions of others into reality while retaining control of the minutiae, would have found it just as easy to talk victory and plot an exit strategy at the same time.

The clues, which double as the countdown to Kammler's disappearance, were there from the beginning of that month.

On 3 April, Kammler had his meeting with Hitler in Berlin. As Goebbels reveals, following their discussions about the 'miracle' weapons that could still win Germany the war, the Führer retained great faith in him. But Kammler had already moved his headquarters out of Berlin to Munich for reasons that Speer guesses at – accurately, it seems – in his memoirs: given the expected failures of these projects, it would have been dangerous for him to remain within Hitler's reach.

Before he left Berlin for the last time, soon after his final meeting with Hitler, Kammler paid Speer a visit. He had come to say goodbye.

'For the first time in our four-year association, Kammler did not display his usual dash,' Speer recalled. 'On the contrary, he seemed insecure and slippery with his vague, obscure hints about why I should transfer to Munich with him.'

Speer might have put this down to the fact that the central part of Kammler's kingdom, at least as far as he, Speer, was aware – the giant Mittelwerk underground A-4 construction complex at Nordhausen in central Germany's Harz Mountains – was about to be overrun by the US First Army. With most of the A-4/V-2 programme team still in place at Nordhausen, Kammler's nervousness was understandable. He needed to get them out.

But then, Kammler told the armaments minister that 'efforts were being made in the SS to get rid of the Führer'.

As Speer was also in the throes of a desperate scheme to assassinate Hitler, using poison gas to kill him and his entourage in the bunker beneath the Reich Chancellery, this perhaps came as no great surprise, except that it came from Kammler's lips. Speer did not know it, but SS General Walther Schellenberg, Germany's combined intelligence chief, had already opened secret surrender negotiations on behalf of Himmler with Count Folke Bernadotte, vice chairman of the Swedish Red Cross. These were transmitted to the Western Allies, but promptly and roundly rejected.

But then Kammler told Speer he was planning to contact the Americans and that in exchange for a guarantee of his freedom he would offer them everything – 'jet planes, as well as the A-4 rocket and other important developments.' He informed Speer that he was assembling all the relevant experts in Upper Bavaria in order to hand them over to US forces.

'He offered me the chance to participate in this operation,' Speer said, 'which would be sure to work out in my favour.' Speer turned him down. In a late display of remorse for his years of unquestioning devotion to Hitler, the armaments minister now saw it as his solemn duty to prevent Hitler carrying out his scorched-earth policy in the face of the advancing Allies. As soon as Hitler was issuing orders for factories to be burned to the ground or blown up, Speer was countermanding them, arguing with frightened and bewildered plant managers that the war was lost and these facilities would be needed for the economic revival of the 'new Germany'.

On 10 April, Army General Walter Dornberger, one-time head of the A-4/V-2 programme, now subordinated to Kammler, ordered 450 key A-4 scientists to evacuate the Nordhausen complex and head for the mountains south of Munich. Those that could not board a special southbound train, nicknamed by those on it the 'Reprisal Express', would be left behind at Bleicherode close by, where Dornberger and von Braun had established their headquarters.

Von Braun argued against this move on the grounds that it would mean deserting his men in the Harz, but Dornberger reminded him that Kammler had issued the order and if they did not obey it, 'the SS would shoot them all.'

Kammler did not come with von Braun and Dornberger on the journey south, but he did make arrangements to meet them at the Hotel Lang, near Oberammergau, which was already serving as an evacuation site for Messerschmitt's advanced projects team, evacuated late the previous year from the company's bombed-out factory at Augsburg.

Von Braun and Dornberger made their way by car to Oberammergau, arriving there on 11 April, the day that US forces entered and overran the underground factory they had just left. The two men, who had co-operated on the rockets from the beginning, were settled into a compound that had been prepared for them by the SS. On his arrival, Von Braun remarked on the beauty of his new surroundings, the snow on the peaks, the lush pasturelands, but noted, too, that the barbed wire had been designed to keep them in, not the Americans out.

Soon after he arrived, von Braun made his way to the Hotel Lang.

As he sat in the lobby, he could hear two people talking in an adjoining room. Straining to catch what they were saying, the rocket engineer recognised the voices as those of Kammler and his chief of staff, SS Obersturmbannführer Starck.

The two men were discussing a plan to burn their uniforms and lie low for a while in the 14th-century abbey at Ettal, a few kilometres down the road. From the tone of the conversation, von Braun was unsure if they were serious or joking. It was at that moment that an SS guard appeared and ushered him into the room.

Kammler, sitting next to Starck, appeared relaxed and affable – very different from the man that Speer had described a week earlier in Berlin.

He asked von Braun how he found the accommodation, whether his team was comfortable, if the broken arm that von Braun had suffered in a car crash the previous month was causing him any great distress, whether the team would be able to begin its design work again soon.

Von Braun, noting the machine-pistol propped beside Starck's chair, replied as enthusiastically as he could to Kammler's questions, careful to maintain the charade. To have done anything less might have been fatal given Kammler's ruthless nature; besides which, von Braun had made provisions of his own.

Before leaving Nordhausen, he had had his men stash ten tons of blueprints relating to the A-4 and its successor, the intercontinental A-9/A-10 rocket, in a deserted iron-ore mine several kilometres from the complex. Had Kammler any inkling of this, von Braun had no doubt as to the fate that awaited them all.

But with the A-4 team settled in Bavaria, the general didn't appear overly concerned about the scientists anymore. He had other matters to attend to, he told von Braun, matters relating to his duties as Hitler's special plenipotentary for jet aircraft production. He would shortly be leaving Oberammergau for an indeterminate period. SS Major Kummer would assume command in his absence. Kammler got to his feet. The meeting was over.

They said their goodbyes and von Braun left.

It was the last time anyone of any standing could corroborate seeing Kammler.

Yet there are traces of his movements in the days, even weeks, that followed; and some of his subsequent actions are decipherable – to a point.

Kammler told von Braun that he had to leave Oberammergau to oversee production of the Me 262 jet fighter according to Hitler's wishes. Yet, five days later, on 16 April, the general appointed Degenkolb to the job, thereby delegating his responsibility; to someone eminently capable, as it turned out. Kammler communicated this decision from his office in Munich, dispatching it to Speer, Goering, Himmler, Hitler's Luftwaffe liaison officer Colonel Nikolaus von Below and SS General Hermann Fegelein, Himmler's liaison officer in the Führerbunker. Describing this communiqué years later as a bafflingly 'unimportant message', even Speer signals his bewilderment at Kammler's determination to spread the word to all and sundry about Degenkolb's appointment.

It seemed obvious with the benefit of hindsight that Kammler wanted as many people as possible to see that he was on top of his assignment – that he was in Munich doing what he had been told to do.

With Degenkolb in charge of Me 262 production, a job that required constant supervision, it left him free to pursue other matters.

When Kammler told Speer that he would offer the Americans 'jet planes and the A-4 rocket', he must have known that their currency was devalued by the sheer numbers of people associated with them. The Me 262, Arado Ar 234 and the Heinkel He 162 jets were in widespread service with the Luftwaffe at the end of the war and plans for hundreds of other jet aircraft were on drawing boards at aircraft factories across the Reich. The Americans and the Russians could take the blueprints for these projects and the design staff associated with them irrespective of Kammler.

The same went for the A-4. As von Braun's actions testify, the rocket had been secretly earmarked for hand-over to the Americans for some time by the people who had designed and built it. Whether Kammler was aware of these moves or not is immaterial; the fact that there were hundreds of V-2s in various stages of construction at Nordhausen meant that the Americans would simply take them as war booty, with only the fate of the design engineers left to negotiate.

If Kammler did have plans to hand the Dornberger/von Braun design team over to the Americans, these came to naught. It appears, though, that he never even tried.

On 13 April, von Braun succeeded in persuading Major Kummer to disperse the scientists into the surrounding villages, ostensibly to cut down the risk of their being wiped out by air attacks on Oberammergau. Von Braun's real fear, of course, was that it would be Kammler issuing the orders for their annihilation.

The move may well have saved their lives, but it also left von Braun free to make his own deal with the US agents that had been dispatched into Bavaria to seek him out.

On 17 April, Kammler sent his message to Himmler about the 'truck' – the last signal he was known to have dispatched from his Munich headquarters. Under German communications procedure, his teletyped 'signature' was preceded with the letters 'GEZ.' – an abbreviation for '*Gezeichnet*' or 'signed by' – denoting to the recipient that he was physically where he claimed to be.

The fact is, he could have been anywhere – anywhere within a narrow corridor of territory then still in the Reich's possession.

As I immersed myself in as many sources as I could regarding Kammler's mercurial existence, a single question pounded repeatedly above the others.

Where had he gone?

Degenkolb was taking care of jet aircraft production. The inner circle in the Führerbunker, cut off from reality and increasingly from the world, had its own problems. No one seemed to care anymore about the whereabouts of the golden boy – the keeper of the miracle weapons who would save Germany from annihilation.

They could have had little idea that he was planning to bargain them for his freedom – for his life.

What did Kammler hope to trade? The rockets were gone. The jets were gone. That left the 'other developments' he had told Speer about.

On 17 April, the GIs of General Devers' 6th Army Group were just days – hours, perhaps – from Munich. All Kammler would have to do was hole up and wait for them.

But what evidence there is suggests he did something quite else.

There is testimony that Kammler moved on from there to Prague, a journey that would have taken him right past the Skoda Works at Pilsen.

Voss had spoken about nuclear propulsion, highly advanced guided weapons and lasers. What other secrets resided within the *Kammlerstab*?

If the witness testimony was accurate, if Kammler did move into Czechoslovakia, what was so precious about the inner workings of the 'special projects group' – this Pandora's Box of advanced weaponry – that it compelled him to go east, not west? Towards the Russians, of all people?

Nothing in Kammler's carefully coordinated life happened by chance. I knew that much. The documentation led nowhere. Even Agoston had reached the end of the trail.

All that was left, was Kammler's lasting legacy: the underground factories that riddled his kingdom, built to his carefully exacting standards, forged by the Holocaust that was also largely of his making.

I requested some time off work and booked myself a fly-drive package to Hanover. From there, with a good hire car, I would be in the Harz Mountains less than two hours after the wheels of the aircraft greased the airport tarmac.

Chapter 17

I entered the complex from the southern end of the Kohnstein – a long ridge dotted with pine and deciduous trees that rose like the hackles of a crouched beast above a sleepy plateau in the Harz Mountains close to the small town of Nordhausen. Inside, I hesitated before drawing breath. Twenty thousand people had died in the construction of Nordhausen, many of them right here, inside the mountain itself.

Anywhere else and you could half-pretend that you were in the bowels of an archaeological site; where lazy reveries about the past of the place were part of the experience. But the imprint of what had happened here within the bare limestone walls was indelible and overwhelming.

I was standing at the entranceway of Gallery 46, a long tunnel with a semi-circular roof. It was still littered with wartime debris. The records said that it had been used in the construction of the V-1 flying bomb.

Every now and again it was possible to recognise a piece of hardware – components of an Argus pulse-jet destined for a V-1, a section of fuel tank or a part of a wing – but, for the most part, in the dim yellow electric light, it could have been any old scrap metal. Next to nothing had been touched, my guide informed me, since this part of the facility had been reopened to visitors in 1994.

Soon after the war, when eventually the Americans left the site – but not before they'd shipped 100 fully assembled A-4/V-2s back to the States – the Soviets sealed Nordhausen by blowing the entrances to the long galleries. In time, the trees had returned to the Kohnstein and East Germans could almost forget that the place had ever been here.

There was a kind of symmetry to this; the Nazis had gone to very elaborate lengths to conceal it during the war, even from their own people.

The Mittelwerk was vast. At the height of its activities, it had been turning out thousands of V-1s and V-2s a month and had consisted of almost 100,000 square metres of floor space. Built in the 1930s as an underground storage facility for fuel-oil, the complex – already huge – gave Speer and Kammler a head start in the construction programme they began and completed between 1943 and 1944.

The site consisted of two main tunnels running in parallel between the northern and southern faces of the ridgeline. One of these, Tunnel B, the main V-2 production line, was a kilometre and a half long.

Between the tunnels, like the rungs of an enormous ladder, were 50 'galleries', each between 100 and 200 metres long. Many had been component manufacturing facilities for the giant rocket assembly line. Some, such as those given over to the manufacture of jet engines, were assembly lines in their own right.

Until the construction of concentration camp Dora nearby, the slave workers who had prepared Nordhausen for the V-1 and V-2 had lived in situ.

Six thousand prisoners had crammed into three tunnels, Number 46 being one of them. Initially, they had slept on straw, their surroundings lit only by a few carbide lamps. Then, they had been allowed to build bunks, four-storey affairs that touched the ceiling, but which still failed to accommodate everyone. Since work went on around the clock, this did not matter; the labourers went to work and slept in shifts.

Beyond Gallery 18, deep inside the mountain, there was no ventilation, water or heating. Every explosion set off by the tunnellers filled the galleries with blinding, choking dust and gases.

This, mixed with the smell of the open latrines – oil-drums sawn in half with two planks on top – made the air almost unbreathable. They also made the tunnels a place of terrible disease and infection. It was a privilege to be on the evening rota that loaded the latrines onto train wagons and emptied them in the open air.

A little more than 50 years later, as I walked the galleries beneath the Kohnstein, I found the silence deeply troubling. I'd wanted to concentrate on the science and engineering, to look at the Mittelwerk as a place that had been responsible for the most awesome piece of technology to have come out of Europe during the war; to breathe the same air as those who'd created the A-4 rocket's assembly line, to soak up that atmosphere in the way that a profiler would use a murder-scene to build a picture of the killer. This place, after all, was one of the final staging posts to Kammler's disappearance. I needed to think as he had thought. But uppermost in my mind was the sheer scale of the crime that had been committed down here and its unexpected proximity to my own ordered existence. The people who'd toiled and died down here to create this Dantean design from Kammler's notebook, the same one perhaps that had spawned his delicate sketches of Auschwitz, were just a generation away from the safe confines of my own world. And all I had succeeded in doing was immersing myself in the fate of its slave-workers, not in the thoughts of the icy technocrat who'd driven them on. In the sub-zero chill of the tunnels, all I could hear were the cries of the dying, not Kammler's whispered escape plans.

Only when I re-emerged from the tunnels could I reapply my mind back to the problem.

When the building work on the tunnels stopped, those that survived went on to work on the rockets. At first, quality control was poor and many of the rockets exploded or went off-course.

Suspecting sabotage, the SS ordered mass executions.

On one day in March 1945, the guards hanged 52 people in Gallery 41, tying a dozen at a time to a beam which was then pulled up by an electrical crane. Those next in line to die were forced to watch.

Then there was Gardelegen, 80 miles to the north. On 13 April, two days after advance US units entered Nordhausen, the SS forced a thousand evacuated prisoners from Dora into a barn that had been pre-soaked with petrol, locked the doors and set light to it. When American troops came across the scene, they found some of the bodies still in the barn, some in a partially dug mass grave. Twenty men had survived by sheltering in holes they had dug beneath the charred corpses of those that had perished in the first moments of the conflagration.

These were Kammler's hallmarks. Others might have pulled the trigger, operated the cranes or poured the petrol, but these acts had been carried out according to his wishes. More than 50 years later, you could sense it.

This was something about the man that I had pulled from the dank, subterranean air.

It was curious, then, that so few people after the war – the many tens of thousands whose lives had been touched by Kammler – could recollect him in any detail. I found this incomprehensible at first, remembering occasions in which concentration camp survivors had identified their tormentors decades after the war.

Even those who had not been directly traumatised by Kammler commented upon this mercurial ability to blend in. A US diplomat who had served in Berlin until the US entered the war in 1941, whom Agoston had interviewed, recalled his ability to 'soothe and tame an unruly horse, using a magically gentle touch, and then minutes later order a negligent groom to be brutally horsewhipped.

'I always thought he was a man to watch,' the American told Agoston. 'But I never saw him again. When I got back to Berlin after the war, no one seemed to remember him.'

How could this man, this monster, the most powerful individual in Nazi Germany outside Hitler's inner circle in the last days of the war, come to be quite so easily forgotten?

It was only outside Nordhausen's giant bomb-proof doors, as I gazed again at the single wartime picture that exists of Kammler – in his general's uniform, striding for the camera, his cap with death's head badge enough to one side to betray more than a hint of vanity – that I began to understand. Kammler was fair, as so many Germans are. There was a gleam of raw intelligence behind those expressionless eyes and a hint of cruelty about the mouth. But take away the uniform, take away the job, the jut to the jaw and the energy in the stride, and he could have been any average 40-year-old European male.

In the chaos of the collapsing Reich, Kammler could have gone anywhere he wanted, assumed any persona he liked, and no one would have been the wiser.

On 11 April 1945, when the first elements of the Third US Armored Division of the First Army swept into Nordhausen, Kammler's legacy would have been plain for the Americans to see.

Clearly, however, his war crimes were not uppermost in their minds.

When Voss was picked up by agents of the US Counter-Intelligence Corps soon after hostilities ended, he told them about the existence of the *Kammlerstab* at Skoda. Their reactions surprised him. They appeared so unmoved by the notion that the special projects group contained a Pandora's Box of exotic military secrets, he concluded that US intelligence already knew.

When Voss went on to propose that they should spare no effort in finding Kammler before the Russians got to him, the CIC agents appeared equally uninterested.

This, from a nation that had mounted the biggest plunder operation of all time – involving the Army, Navy and Air Force, as well as civilian agents – to bring advanced German technology back to the United States.

At first, I wondered if their behaviour had to do with the fact that Pilsen had fallen within the designated Russian zone of occupation. Maybe the CIC recognised that Kammler was beyond their reach and had simply abandoned any hope of finding him?

Then I came upon the final piece of the jigsaw.

The rapid eastward push that had brought the Third Armored Division to Nordhausen did not stop there. Disregarding agreements signed by the exiled Czech government and the Soviet Union, troops of Patton's Third Army to the east of Nordhausen crossed the Czech frontier on 6 May.

Deep into the Soviet-designated zone of occupation, a forward unit of Patton's forces entered Pilsen that morning. Records I'd had sent to me from the US National Archives before I left England showed that US forces had the run of the Skoda Works for six days, until the Red Army showed up on 12 May. Following protests from Moscow, the US Third Army was eventually forced to withdraw.

Six days is a long time if you're retrieving something you already know to be there.

Had Kammler already done his deal with the Americans? Speer, during his initial interrogation at the hands of the US Strategic Bombing Survey Mission on 21 May, less than two weeks after the German surrender, intuitively believed so.

Asked by the mission for technical details of the V-2, he replied: 'Ask Kammler. He has all the facts.'

As I pushed eastwards towards the Czech border, using wartime documents I'd gathered in London and a worn copy of *Blunder!* as my guide, I found myself following a scent of collusion that was still choking more than 50 years later.

Chapter 18

On the morning of Sunday 6 May 1945, five days after the announcement of Hitler's death in Berlin, troops of the US 16th Armored Division pushed cautiously through an eerie, soulless landscape that its soldiers described as 'neither German nor Czech'. The local people – German-speaking, terrified by their Slav neighbours to the east and undecided whether to remain in their homelands or flee westwards – were powerless to resist the US troops.

The war was all but over and everyone knew it; everyone except a hard core of Waffen-SS in and around Prague, southern Czechoslovakia and Austria that had signalled their intention to defend this, one of the final remaining pockets of the Reich, to the very last.

As the half-tracks rolled deeper into Czech territory, the US soldiers passed ragged columns of Wehrmacht troops rolling westwards to the comparative safety of the US zone of occupation. The US objective was Pilsen, a place known to the soldiers of the 16th Armored for its beer, not the giant armaments factory where Skoda had cranked out guns and munitions for the Nazis for the best part of seven years.

Inside the US half-tracks, you could cut the atmosphere with a knife. No one wanted to end up a statistic in the last few days of the war. But still, the pace was relentless. Rumour had it that Patton wanted to reach Prague ahead of the Russians. And that meant a confrontation with the SS.

Approaching Pilsen, the atmosphere changed. The US troops noticed a scattering of red, blue and white Czech flags in place of white flags of surrender. As the half-tracks left the Sudetenland behind and pushed on into the heartland of Czechoslovakia proper, the people came out into the streets, waving and cheering as the 16th Armored swept past. 'It was Paris all over again, with the same jubilant faces, the same delirium of liberation,' one soldier recalled in the 16th Armored Division's official account of the liberation of Pilsen.

By 0800 hours, the US troops were in the centre of Pilsen, their advance all but halted by the crowds that had spilled off the pavements and into the streets. The situation was chaotic. It was only when German snipers opened up in many sections of the city an hour after the Americans arrived that sanity of a kind was restored. The civilians fled back into their homes, leaving the US soldiers to mop up what resistance remained and carry out house-to-house searches.

One unit from the 16th Armored soon came upon Lt Gen Georg von Majewski, the German army commandant in Pilsen, with 35 members of his staff in the Wehrmacht's town headquarters. As American troops entered von Majewski's office, the general surrendered his forces to the officer in charge, then pulled a pistol from the drawer of his desk and shot himself through the head in front of his wife.

Von Majewski was one of 14 enemy dead that day. Around 4,500 more surrendered. The Americans had suffered one fatal casualty in the whole Pilsen operation.

A second 16th Armored unit, Task Force 'Casper', captured the airport and a third, Task Force 'Able', occupied the Skoda Works.

An air raid by US B-17s several weeks previously had rained destruction on a large part of the facility. The forward units of Task Force Able reported that the central administrative building had been 80 per cent destroyed.

The news that Pilsen had fallen to the Americans was picked up on the radio that night by the one-time head of the Skoda Works, Dr Wilhelm Voss, who was hiding out with his wife and family at a hunting lodge on a Skoda estate south of Prague.

Voss had been out of a job since January, when Goering had dismissed him for refusing to accept two of his nominees on the Skoda main board.

For weeks, Voss had been monitoring the Allies' radio reports for news of their advance, knowing that it was touch and go whose forces would get to Pilsen first. Armed with the news he wanted to hear, Voss seized the moment soon after the German surrender on 7 May and set out for Skoda.

Given Voss' former position and the chaos that engulfed the former Protectorate, this was a hazardous undertaking. Czech partisans were summarily executing all the Germans they could find. While the bulk of the Waffen-SS units had officially surrendered, some were reported still to be holding out against the Russians in and around Prague. But as Voss told Tom Agoston during their conversations in 1949, he was determined to hand over the blueprints of Kammler's SS special projects group to the Americans, not the Russians.

According to *Blunder!*, Voss finally pulled up in front of Skoda's main gates on 10 May. He bluffed his way past the American guard and found a very different place from the one he remembered; a Skoda in transition, with Germans, Czechs and Americans milling about the plant. In the confusion, he located a member of the *Kammlerstab*, who told him that survivors from the group had already recovered some documents, marked them up as personnel and payroll records, and loaded them onto a truck. They were hoping to get them to a place of safekeeping before the Russians arrived. The facility was due for hand-over to the Red Army in two days' time.

Sensing that this was the time to act, Voss sought out the US officer in charge, informed him who he was, then told him what was in the truck. The American listened, unimpresed by what he was hearing, and finally informed Voss he was under orders to hand everything over to the Russians. Voss insisted, but the American was adamant; nothing was to be moved. An American ordnance team had already inspected the plant earlier that week, he told the exasperated industrialist, and had picked up everything it needed.

Two days later, Voss watched as the truck was driven away by a Red Army transport officer.

In *Blunder!*, Tom Agoston had traced records showing that a British-led Combined Intelligence Objectives Sub-Committee team entered the Skoda Works on 12 May and promptly got the runaround from Skoda's newly ensconced Czech managers, who wanted to cooperate with the Russians and no one else.

I had found another set of CIOS reports, from a British team that entered Skoda one week later on the 19 May with the plant now firmly under Soviet jurisdiction. This report, CIOS Trip No. 243, summarised a combined US-UK operation – the US team, it said, having entered the plant on 16 May – in which nothing of any real interest was discovered, except for certain new projects in the ammunition field and some advanced German design and production methods for guns and munitions.

The triple-ring of security put in place by Himmler and Kammler around the *Kammlerstab* had done its job, even after hostilities had ended.

Voss' story implies that Kammler never made it to the Skoda Works to retrieve the blueprints of the *Kammlerstab* – that instead the papers were ignominiously removed by the Russians and taken back to the Soviet Union.

It didn't look that way to me.

Based on what was happening in Germany, where US technical intelligence agents were systematically targeting any weapons facility of interest and plundering the contents, it is inconceivable that they would have ignored Skoda, even if they had been unaware of the SS-run special projects group in its midst.

The US officer who stonewalled Voss over the papers in the truck told him that a US 'ordnance team' had inspected the plant earlier that week. 'Earlier that week' could only mean that the 'ordnance team' had walked into the plant alongside or directly behind the combat units of 16th Armored's Task Force Able on 6 May.

There is no official record of any such unit's visit to Skoda. There are, however, reports that a US team with nuclear expertise entered the plant and gained access to documents that outlined the SS cell's work in conjunction with the Junkers company at Dessau on nuclear propulsion for aircraft. I had been informed by someone that an account of this activity was contained in *Atomic Shield*, the official history of the US Atomic Energy Commission. But after extensive enquiries with archivists at the US Department of Energy, the AEC's successor, I was assured that there were no references in that book or any other to Skoda, its purported SS think-tank or German nuclear research at Pilsen or Dessau.

Given Voss' claims that nuclear propulsion work had been one of three primary areas of research within the *Kammlerstab*, alongside lasers and guided weapons, it is inherently credible that some such work had gone on there.

It is at odds, however, with the official view that Germany was nowhere near as advanced in nuclear weapons technology as the Americans, who by 10 May were less than two months away from detonating their first atomic bomb in the New Mexico desert.

According to Speer, official Nazi sanction for attempts by Werner Heisenberg, the father of the would-be German atomic bomb, were rescinded in 1942 on the grounds that the earliest timetable for an available weapon was 1946 – too late, Speer concluded, to be of any use in the war.

In late 1944 and early 1945, US technical intelligence agents from the Alsos Mission ('alsos', in a quirky piece of coding, being ancient Greek for 'grove', i.e. General Leslie Groves, who was in charge of the US bomb project) tore into France, Belgium and Germany searching for signs of German atomic bomb work and quantities of German-held uranium oxide ore – a material that is inert until it has been subjected to neutron bombardment in a nuclear reactor.

On 23 April, an Alsos team led by the unit's commander, Lt Col Boris T. Pash, a former high school teacher turned Army G-2 security officer trained by the FBI, arrived at the locked steel door of a box-like concrete entranceway in the side of a cliff above the picturesque German town of Haigerloch, near Stuttgart.

Pash, who had been in a race against the French to get there, shot off the lock and found himself in a chamber with a concrete pit about ten feet in diameter in the middle. Within the pit hung a heavy metal shield covering the top of a thick metal cylinder. The latter contained a pot-shaped vessel, also of heavy metal, about four feet below the floor level.

What Pash had discovered is what history describes as one of only two German reactors assembled, or partially assembled, before the end of the war – the other, at Kummersdorf, being captured by the Russians.

Using heavy water as a moderator and 664 cubes of metallic uranium as its fuel, the Haigerloch reactor had achieved a seven-fold neutron multiplication just a few weeks earlier.

Heisenberg had calculated that a 50 per cent increase in the size of the reactor would produce a sustained nuclear reaction.

Had he done so, Hitler would have had the means to enrich enough fissionable material to construct a bomb.

This is the official view of the extent of the German atomic bomb programme and it is terrifying enough. The Nazis had come close to developing a weapon.

But look a little to the left and right of the official view, as I had before· leaving London, and an even more frightening scenario pulled into view.

On 19 May, a German U-boat, the U-234, docked at the US Navy port of Portsmouth, New Hampshire, having surrendered to a US Navy destroyer two days earlier off the eastern seaboard. Her commander had received high-level orders to sail from Kristiansand in Nazi-held Norway to Japan on 16 April.

On board, the Americans found an Aladdin's Cave of technical equipment and blueprints, most of them related to advanced German jet aircraft. The U-234 was also ferrying technical experts, a nuclear technician amongst them.

This fact alone tipped off US investigators that U-234 was no ordinary U-boat. A secondary examination showed she had been modified to carry a very dangerous cargo. Set in her six adapted mine-laying tubes were 560 kilos of uranium held in ten gold-lined containers. The loading manifest maintained that it was uranium oxide, the state in which uranium is found when it is extracted from the earth and safe enough to carry around in a paper bag. That the U-234's converted mine-laying tubes were gold-lined indicates that its cargo was emitting gamma radiation. This, in turn, meant the uranium oxide ore had been subjected to enrichment from a working nuclear reactor. The gold-lining was to prevent the radiation from penetrating the U-boat's hull.

And yet, officially, there had been no nuclear reactor in Germany capable of fulfilling this task.

Not in Speer's orbit of operations. But what about Kammler's?

The journey to this point had revealed the extent to which the SS had ring-fenced its top-secret weapons activity from the rest of the German armaments industry. Could the *Kammlerstab* have contained the plans of a working nuclear reactor? Its secret location, even? It was a possibility I had to confront.

The lack of data in US archives concerning the fate of the U-234, its cargo and its personnel is astonishing. But then, everywhere I'd looked prior to leaving London, a pattern had been emerging.

Protracted searches by archivists at the US National Archives for any data on Kammler had failed to locate a single entry for him.

It was the same for Skoda.

Given Kammler's range of responsibilities in the final months of the war, this absence of evidence was remarkable; so much so, that one archivist at Modern Military Records, College Park, Maryland, said that it 'redlined' him in her book.

I asked what she meant.

Somebody, she replied matter of factly, had been in and cleaned up.

Tom Agoston had found evidence of much the same thing. Despite Voss' extensive debriefings on Skoda by US CIC agents at the US European Command Counter-Intelligence Center at Camp Oberursel, near Frankfurt, no one had been able to trace any records or transcripts of the interrogation.

When Agoston ran checks on Kammler, it transpired that his name hadn't even registered at the Nuremberg war crimes trials – the very least that might have been expected of someone who had played such a prominent role in the Holocaust.

Unlike Martin Bormann, whose body was never found after he was reported killed during his escape from the Führerbunker on 1 May, Kammler was never tried *in absentia* at Nuremberg. There is no evidence that anyone ever bothered to look for him. This, despite the fact that the four versions of his death conflict with each other so markedly that none stacks up under scrutiny.

The first version says he committed suicide in a forest between Prague and Pilsen on 9 May, two days after the German unconditional surrender.

The second says that he died in a hail of bullets as he emerged from the cellar of a bombed-out building to confront Czech partisans that same day.

The third version says that he shot himself in a wood near Karlsbad (Karlovy Vary in today's Czech Republic), also on 9 May.

What these variations on a theme do manage to assert is that Kammler was in or around Prague just prior to the German capitulation.

One witness quoted by Agoston, an official with the Prague regional office of the Buildings and Works Division of the SS Economic Main Office, recalled the date: 'Kammler arrived in Prague early May. We did not expect him. He gave no advance notice of his arrival. Nobody knew why he had come to Prague, when the Red Army was closing in.'

I could see only two reasons why Kammler should have made such a journey. First to retrieve the mother-lode of documentation on the special projects group at Skoda and the Skoda administrative offices in Prague; and secondly to bury it somewhere prior to setting up the deal that would save him from the gallows and buy him his freedom.

Where in the crumbling, collapsing edifice of the Reich would Kammler have considered hiding this priceless cargo? Certainly not in Czechoslovakia, not with the Russians closing in. And it was unlikely that he would have headed west through the partisan-infested forests between Prague and the German border.

The fourth version of his death, I felt, offered the best possible clue.

The German and Austrian Red Cross, according to Agoston, held two relevant documents after the end of the war, the first posting Kammler as 'missing', the second updating the first and listing him 'dead'. The first report had been filed by a relative. It placed Kammler's last known location as Ebensee in Steiermark, Austria. The updated report was based on the testimony of unnamed 'comrades'. No burial place was specified.

I did some thinking of my own, using the reference books I had brought with me as pointers.

In early May, the province of Steiermark was putting up a vigorous defence against the Russians, whilst bracing iself for the American advance, under the leadership of Gauleiter Siegfried Ueberreither. The gauleiter – the Nazi party's chief administrator for the Steiermark 'gau' – was backed in his decision to hold the province to the last by large numbers of Wehrmacht and SS troops.

Kammler knew Ebensee well. Under his orders and supervision, work had commenced there in 1943 on a giant underground complex for the intercontinental A-9/A-10 rocket – the weapon that Hitler had intended to use against New York. The facility was codenamed Zement and a large section of it had been completed by the end of the war. Zement was tucked into the mountains abutting the shoreline of Lake Traunsee.

On 8 May 1945, Lake Traunsee could be considered the epicentre of the Nazi empire on the eve of its surrender; a seething confluence of terrified citizenry, refugees from the east, heavily armed German troops and Nazi party members looking for places to lie low until the frenzy of capitulation had subsided. Gauleiter Ueberreither himself, having urged his people to fight to the last bullet, slipped out of his office in Graz and disappeared. Years later, he emerged in the Argentine.

I began to make arrangements that would get me to Ebensee. But shortly before I could complete them, I received a call from Poland. A researcher there had responded to my request for information about German wartime research activities around Breslau, one of the two purported homes of German flying saucer research.

The guy's name was Igor Witkowksi and he had been recommended to me by Polish sources through my work at Jane's as someone who was both highly knowledgeable and reliable. Witkowski, they told me, had self-published a series of books in Polish detailing his findings.

For years, Witkowski had been sifting through archives related to the many secret Nazi scientific establishments in and around Breslau – modern-day Wroclaw – looking for traces of wartime secret weapons work. What he'd uncovered was evidence of a systematic SS-controlled R&D operation in a rat-run of interconnected facilities, many of them underground, in the remote Sudeten Mountains in the south-west of Poland – Lower Silesia as it had been, before its hand-over to Poland under post-war reparations by Germany.

Witkowski's research led him to conclude that a number of these facilities had been used by the SS for nuclear research work. But one site did not conform to the pattern. In this case, he said, a series of experiments had taken place in a mine in a valley close to the Czech frontier. They had begun in 1944 and carried on into the April of the following year, under the nose of the advancing Russians.

The experiments required large doses of electricity fed via thick cabling into a chamber hundreds of metres below ground. In this chamber, a bell-shaped device comprising two contra-rotating cylinders filled with mercury, or something like it, had emitted a strange pale blue light. A number of scientists who had been exposed to the device during these experiments suffered terrible side effects; five were said to have died as a result.

Word had it that the tests sought to investigate some kind of anti-gravitational effect, Witkowski said. He wasn't in a hurry to agree with this assessment, he added, but he was sufficiently intrigued by the data to alert me to the possibilities.

Like Witkowski, I was sceptical, a feeling that grew when he told me who had presided over the work: Walther Gerlach, Professor of Physics at the University of Munich and head of nuclear research at the Reich Research Council in 1944–45.

Wilhelm Voss, joint head of the SS special projects groups inside the Skoda Works, had made it clear to Tom Agoston that German nuclear work had absorbed a large part of the group's research activities. By early May, Kammler should have had adequate intelligence that Pash, the FBI-trained former teacher and head of the US Alsos Mission, had located most, if not all Germany's nuclear research centres in the western zone of influence along with the scientists that went with them. Kammler might also have known that the U-234 had sailed with its cargo of uranium ore for Japan; maybe he'd been responsible for the order himself.

While he could not have known that the Americans were close to perfecting their own nuclear weapon, or that the U-234 would fall into US hands, he didn't seem like a man who would gamble his life on providing America with a technology it had already acquired via diligent detective work. It didn't tally with his profile.

My feeling was that Kammler would offer them something so spectacular they'd have no choice but to enter into negotiations with him; and, intriguing as it was, this experiment down the mine didn't seem to fit the bill. The sickness suffered by the scientists appeared to be the result of radiation poisoning. As for the device itself, I said, by now thinking out loud, since it fell within the Soviet zone of influence, it hardly conformed to the plan that Kammler had outlined to Speer in Berlin.

It was then that Witkowski, who had listened for several minutes patiently enough, interrupted me. His research had uncovered the existence of an SS-run 'Special Evacuation *Kommando* (command, team or unit)' that had evacuated the 'Bell' and its supporting documentation prior to the Russians' overrunning the facility. The evidence said it had been shipped out, destination God only knew where; except it wasn't there when the Russians arrived.

Despite the warmth of my hotel room, I felt a sudden chill. For all the months I had been following Kammler, he had been a remote figure, even at his closest in the wet, freezing-cold underground galleries at Nordhausen. Now, fleetingly, I felt him in the same room as me.

'And the scientists?' I asked. 'What happened to them?'

I thought the line had gone dead, but it was just Witkowski taking his time, choosing his words.

'They were taken out and shot by the SS between the 28th April and the 4th May 1945. Records show that there were 62 of them, many of them Germans. There were no survivors, but then that's hardly surprising.'

He paused. 'It's quite clear that someone had gone to very great lengths to clean up.'

It was the second time in as many weeks that this expression had been used in the context of Kammler. Within the hour, I had booked myself on a flight to Warsaw.

It was thus – simply and irrevocably – that I was pulled into the extraordinary story of the Bell.

Chapter 19

We left the Polish capital in the haze of an overcast dawn, moving quickly past drab living blocks half-hidden by a frenzy of construction work.

Witkowski drove, apologising in advance for the schedule ahead. The Wenceslas Mine was 50 kilometres or more the other side of Wroclaw and Wroclaw, formerly Breslau of the German *gau* of Lower Silesia, was around 350 km from Warsaw. The drive would be hard, he said, much of the route consisting of poorly maintained single-lane roads, but all being well we would reach our destination before sunset. In the meantime, he would bring me up to speed on his research, which had started with a phone call from a Polish government official.

Witkowski, a former defence journalist, did not want to tag his informant, but he assured me that his integrity was impeccable. The official had permitted him to view some documents, to make notes, but not to take anything away with him.

Witkowski had taken him at his word and transcribed most of what he had been allowed to view verbatim.

At the airport, his stocky appearance was just as I had imagined him. He was younger than me, with a thick-set expression, high Slavic cheek-bones and dark, intensely serious eyes. From the word go, we talked about little else but the research that united us. Had Witkowski been in any way a lightweight, I would have turned around and got on the first plane home. But when I saw him, I knew he was OK. There were aspects to him that I saw in myself. Like me, Witkowski was consumed by a thirst for knowledge. Knowledge of a highly specialised kind.

After months on the road it felt good to be sharing data again. And the very fact he'd let me in on his discovery told me that Witkowski felt the same way. We were treading a tightrope. If either of us were wrong about the data we'd collated, it'd be a long fall. On the phone, Witkowski had convinced me that he could take me to places that only he and a select group of local researchers knew about.

The papers that Witkowksi had been allowed to transcribe detailed the activities of a special unit of the Soviet secret intelligence service, the

185

NKVD, forerunner of the KGB. Attached to this unit were two Polish officers, General Jakub Prawin, head of the Polish military mission in Berlin, and Colonel Wladyslaw Szymanski, a senior member of his staff.

It was during their 'debriefing' of a former high-ranking official from the *Reichssicherheitshauptamt*, the Reich Central Security Office or RSHA, that Prawin and Szymanski learned of 'General Plan 1945'.

Their interrogation subject, Rudolf Schuster, had worked at the RSHA – one of four security organisations run by the SS-controlled *Sicherheitsdienst* or Security Service – until his deployment on 4 June 1944 to the Special Evacuation *Kommando*, an organisation unknown to the NKVD, which prided itself on knowing everything. Schuster, it transpired, had been responsible for the ELF's transportation arrangements.

Schuster's information must have set alarm bells ringing at the highest Soviet levels, for Prawin and Szymanski had learned that the man behind General Plan 1945 was none other than Martin Bormann, Hitler's deputy, last seen fleeing the Führerbunker on the night of 1 May. Though supposedly killed in the attempted break-out, Bormann's body had never been formally identified. Tying Bormann into a scheme for some kind of evacuation procedure, therefore, had disturbing implications, but pumping Schuster was evidently pointless as it was clear he was only privy to the transportation details. The SS, operating its traditional compartmentalisation system of security, had kept him from the broader aspects of the plan. It was apparent, however, that much of the *Kommando*'s activities centred on the territory of Lower Silesia, now southwest Poland.

The next piece of the jigsaw was provided by SS Obergruppenführer Jakob Sporrenberg, who from 28 June 1944 had commanded a section of the *Kommando* attached to the Gauleiter of Lower Silesia, Karl Hanke. Hanke, like all gauleiters, who were the party's representatives in the gaus or regions, was answerable directly to Bormann, the party chief.

Sporrenberg had been captured by the British in May 1945, but was handed over to the Poles because he was a general in the SS and had been active in Poland, the scene of many of the SS-run death camps. Had they realised Sporrenberg's true brief, it is unlikely they would have been quite so quick to release him.

Sporrenberg was a big shot, one of a scattering of senior SS police generals of exalted rank and status positioned to give Himmler absolute power throughout Germany in the closing years of the war. To put his seniority in perspective, he carried the same rank as Ernst Kaltenbrunner, the brutal Austrian who succeeded Heydrich as head of the Reich Security Office, the Nazis' central agency for internal counter-espionage and repression. For Sporrenberg to have been assigned the command of the Special Evacuation *Kommando* unit attached specifically to Hanke's gau of Lower Silesia shows how important Bormann's evacuation plan was considered to be.

And, I thought, just how fucking secret. There has never been any official acknowledgement of the existence of the Special Evacuation *Kommando*.

Exactly what, I asked Witkowski, was the *Kommando* tasked with evacuating?

The taciturn Pole, concentrating on the road ahead, flashed me a look. Patience. Everything in its own time.

Sporrenberg, he continued, was sent for trial and sentenced to death in 1952, but not before he had testified in secret before the Polish courts about his role as one of Lower Silesia's main plenipotentaries for the evacuation of high-grade technology, documents and personnel or his part in the murder of 62 scientists and lab-workers associated with a top-secret SS-run project in a mine near Ludwigsdorf, a village in the hills south-east of Waldenburg, close to the Czech border.

Sporrenberg, operating under the political oversight of Gauleiter Hanke, had been in charge of a *Kommando* cell tasked with 'northern route' evacuations via Norway, which remained in German hands until the very end of the war.

The NKVD/Polish intelligence team learned that SS-Obersturm-bannführer Otto Neumann, who had commanded a *Kommando* detachment in Breslau, had been responsible for all south-bound shipments to Spain and South America.

Neumann was never caught, but was reputed to have escaped to Rhodesia, where he was supposedly sighted after the war.

According to some estimates, Witkowski said, the air-bridge established by the *Kommando*'s southern command between the Reich's remaining occupied territories and neutral, but Axis-sympathetic Spain managed to evacuate 12,000 tonnes of high-tech equipment and documentation in the final months of the war using any commandeered Luftwaffe transport it could lay its hands on.

The *Kommando* southern command had one other exit route available to it up to the end of the war – albeit an even more hazardous one. Using some of the northern Adriatic ports that remained in German hands up to the surrender, a brave or foolish U-boat commander could conceivably have run the gauntlet of Allied air and maritime superiority to evacuate cargo and personnel by sea.

But it was Sporrenberg's northern route operations that underpinned the reason for the trip, because it was Sporrenberg who provided the only details that have ever come to light on the Bell.

The experiment started out at a top-secret SS-run facility near Leubus (Lubiaz in modern-day Poland), northwest of Breslau, in early-mid 1944. With the Soviets' rapid push into Poland during late 1944/early 1945, the unit was transferred to a castle on a hill above the ancient village of Fuerstenstein (Ksiaz), 45 km to the south, close to the Lower Silesian coal-mining centre of Waldenburg. From there, it was moved again to the mine near Ludwigsdorf (Ludwikowice), 20 km the other side of Waldenburg, nestling in the northern reaches of the Sudeten Mountains. The Wenceslas Mine, where the Bell ended up, had been requisitioned by the SS as part of a neighbouring underground weapons complex, codenamed *Riese* – 'Giant'.

Riese, only part-completed by the end of the war, was an attempt to transform an entire mountain into an underground weapons production centre. The many tens of kilometres of galleries that had been tunnelled by the end of the war had been clawed from the rock by inmates drafted in from the nearby concentration camp of Gross-Rosen. Modern excavations of *Riese* show that the SS had been attempting to link it to the Wenceslas mine via a tunnel almost ten kilometres long.

By mid-afternoon, Witkowski and I had reached Wroclaw, a smoky industrial city on the banks of the River Oder that in 1945, as the 'fortress' city of Breslau, had held out against the Red Army for more than 70 days, despite total encirclement.

In September 1944, long before the Russians were at Breslau's gates, the German military commander, General Krause, proposed the evacuation of 200,000 civilians, but Gauleiter Hanke refused, seeing it as a sign of weakness.

Four months later, as the Soviet ring of steel closed around the city, Hanke finally relented and women and children were permitted to leave.

But in the depths of a winter that was cold enough to freeze the sweeping waters of the Oder those that weren't cut down, killed or raped by the Russians succumbed in their thousands to the sub-zero temperatures.

In February, the troops of SS Fortress Battalion 'Besselein' launched a ferocious counter-attack in an attempt to break the siege. While the German troops met the Russians head-on in some of the heaviest street-fighting since Stalingrad, Hanke exhorted his citizens to victory under the slogan 'Every house a fortress'. He then ordered a ruthless conscription of manpower that produced five regiments of Volkssturm –

old men beyond normal fighting age – and Hitler Youth.

Following the briefest of training regimes, principally in the use of anti-tank 'panzerfausts', they went out and died in their thousands.

We edged forward in the heavy traffic, soon finding ourselves on a long wide thoroughfare, the former Kaiserstrasse, which by April 1945 had been transformed into a landing strip for medium transport aircraft after the Russians captured the airport. It signalled the beginning of the end, but still the defenders fought on.

German Red Cross flights of Junkers Ju 52s removed 6,000 of the more seriously wounded and beneath the dense overcast that merged with the factory smoke a half century later, I could almost hear the rumble of tri-motors as they swept into land, touching down at night under the guiding light of flame-torches, and the crump-wump of artillery explosions in the suburbs.

By this stage, the defenders – soldiers and civilians alike – were forced to battle the Russians on all conceivable fronts, including the sewers beneath the streets.

On 4 May, two days after the Russians took Berlin, a delegation of churchmen begged General Niehoff, Krause's replacement, to surrender. Two days later, he agreed and on 7 May the Red Army marched in.

Furious at having been held at bay by such a vastly inferior force, the Soviets launched heavy reprisals against the soldiers and citizenry of Breslau; a fate that Gauleiter Hanke chose not to share. He had flown out of the city on 4 May in a Fieseler Storch liaison aircraft to take up his post as successor to the disgraced Himmler, whose surrender negotiations had been uncovered by Hitler shortly before his suicide.

Hanke, local administrator of Bormann's enigmatic General Plan 1945, the man to whom the Breslau-based detachment of the Special Evacuation *Kommando* was directly acountable, was never seen again.

When I told Witkowski about Kammler, how he too had vanished off the face of the earth, he was unmoved. It was simply part of a pattern, he told me. Hanke, Kammler . . . there were dozens of high-ranking former SS or Party members that had never been called to account. They had simply disappeared. Many of them shared the distinction of having had access to highly advanced technology.

In the immediate aftermath of his death sentence, Sporrenberg, head of the Special Evacuation unit's Breslau operations, was smuggled to Russia, milked by the NKVD in one last brutal round of interrogation, then disposed of. Schuster, in charge of the unit's transportation arrangements, died 'mysteriously' in 1947 and Szymanski and Prawin, the two Polish intelligence officers who'd interrogated them, also met untimely ends after their release from the special NKVD investigation cell – Szymanski in an air crash and Prawin in a drowning accident.

'What was it that killed them?' I asked. 'Their knowledge of the Special Evacuation *Kommando* or the fact they knew about the mine?'

Witkowski kept his eyes on the road. 'The Germans maintained a special detachment of aircraft, probably Ju 290s, and a single Ju 390, both rare heavy transport types, based at Opeln – Opole today – a hundred kilometres from here. Witnesses say the planes were well camouflaged and that some were painted with yellow and blue identification markings, which suggests they were masquerading as Swedish planes. If so, it was almost certainly a unit of KG 200, the Luftwaffe special operations wing, which frequently flew aircraft under the flag of enemy or neutral countries. The point is, the Evacuation *Kommando* had the ability to move thousands of tons of documentation, equipment and personnel, and they could move it north or south. Sporrenberg's brief was to go north. The whole operation was highly secret. But nothing, it seems, was more secret than the Bell.'

We crawled out of Wroclaw in a heavy convoy of trucks bound for the Czech border. I focused on a faint ribbon of sky between the clouds gathering to the south and the outline of the Sudeten Mountains, the inner core of Kammler's high-tech kingdom, indistinct yet somehow threatening, on the distant horizon.

Chapter 20

In his interrogation sessions with the NKVD and his deposition to the Polish courts, SS General Jakob Sporrenberg could only tell the interrogators and the prosecutors so much, Witkowski told me. Sporrenberg was a policeman and an administrator, not a technician. He could relay what he had seen and heard, little snippets he had picked up, but the technology itself was beyond him. He had been tasked with getting men, materiel and documents out of Lower Silesia before the Russians reached them; he had had no need to know much else. This was standard SS operating procedure.

I asked what else he knew about Sporrenberg.

Witkowski stared straight ahead as he talked. First, I had to understand that Sporrenberg was very high up the chain of command. Records showed, he said, that Sporrenberg was appointed deputy commander of the Waffen-SS VI Korps under SS-Obergruppenführer Walter Krueger in 1944. Krueger, as far as Witkowski had been able to ascertain, had been involved in a series of top-secret SS-conceived operations in the closing months of the war, including a plan to occupy neutral Sweden, the evacuation of Nazi wealth to South America and other neutral or non-aligned states, a hare-brained scheme to strike at New York with V-1s launched by submarine and, of course, the special evacuation of Nazi secret weapons and high-technology. Sporrenberg was appointed plenipotentiary in charge of northern-route 'special evacuations' to Norway on 28 June 1944. Like Kammler, it seemed that General Sporrenberg was recognised and rewarded for his ability to organise, not for his ability to fight.

Following his capture, as much as Sporrenberg was able to divulge to Soviet intelligence and the Polish courts about the Bell was this, Witkowski said. The project had gone under two codenames: '*Laternentrager*' and '*Chronos*' and always involved '*Die Glocke*' – the bell-shaped object that had glowed when under test. The Bell itself was made out of a hard, heavy metal and was filled with a mercury-like substance, violet in colour. This metallic liquid was stored in a tall thin thermos flask a metre high encased in lead three centimetres thick.

The experiments always took place under a thick ceramic cover and involved the rapid spinning of two cylinders in opposite directions. The mercury-like substance was codenamed 'Xerum 525'. Other substances used included thorium and beryllium peroxides, codenamed *Leichtmetall*.

The chamber in which the experiments took place was situated in a gallery deep below ground. It had a floor area of approximately 30 square metres and its walls were covered with ceramic tiles with an overlay of thick rubber matting. After approximately ten tests, the room was dismantled and its component parts destroyed. Only the Bell itself was preserved. The rubber mats were replaced every two to three experiments and were disposed of in a special furnace.

Each test lasted for approximately one minute. During this period, while the Bell emitted its pale blue glow, personnel were kept 150 to 200 metres from it. Electrical equipment anywhere within this radius would usually short-circuit or break down. Afterwards, the room was doused for up to 45 minutes with a liquid that appeared to be brine. The men who performed this task were concentration camp prisoners from Gross-Rosen.

During the tests, the scientists placed various types of plants, animals and animal tissues in the Bell's sphere of influence. In the initial test period from November to December 1944, almost all the samples were destroyed. A crystalline substance formed within the tissues, destroying them from the inside; liquids, including blood, gelled and separated into clearly distilled fractions.

Plants exposed to the Bell included mosses, ferns, fungi and moulds; animal tissues included egg white, blood, meat and milk; the animals themselves ranged from insects and snails to lizards, frogs, mice and rats.

With the plants, chlorophyll was observed to decompose or disappear, turning the plants white four to five hours after the experiment. Within eight to fourteen hours, rapid decay set in, but it differed from normal decomposition in that there was no accompanying smell. By the end of this period, the plants had usually decomposed into a substance that had the consistency of axle grease.

In a second series of experiments that started in January 1945, the damage to the test subjects was reduced to around 12–15 per cent following certain modifications to the equipment. This was reduced to two to three per cent after a second set of refinements. People exposed to the programme complained of ailments, in spite of their protective clothing. These ranged from sleep problems, loss of memory and balance, muscle spasms and a permanent and unpleasant metallic taste in the mouth. The first team was said to have been disbanded as a result of the deaths of five of the seven scientists involved.

This, Witkowski said, was all contained in the documents he had been shown. What impressed him – and what I now confessed had me intrigued as well – was their level of apparent detail. It told him – and me – that in all likelihood *something* had happened down the mine; something mysterious. But what? Witkowski insisted that it had all the hallmarks of an anti-gravity experiment. But I wasn't so sure. Before I could begin to believe this was the Holy Grail, I needed to tick off a set of more rational explanations first. And the most glaring of these was the probability that this had been some kind of test involving nuclear material. The documents that Witkowski had seen had mentioned the involvement of Professor Walther Gerlach, the man charged with oversight of Germany's atomic weapons programmes. Disturbingly, they also cited Dr Ernst Grawitz, head of the euphemistically labelled SS Medical Service. Grawitz had been the boss of Josef Mengele, the infamous doctor of Auschwitz. Inevitably, there were reports that the Bell had been tested on humans as well. Again, if the Bell had emitted radiation, the Nazis – given their record on war crimes in this area – would have probably monitored its effects.

I said nothing, however, out of deference to Witkowski, who told me that much of the rest of his evidence was visible at the Wenceslas Mine itself.

As we approached Waldenburg, Walbrzych as it is today, the landscape altered. After nothing but lowlands and flatlands all day, the road now started to twist and climb as we headed into the foothills of the Sudeten range.

Waldenburg itself was bleak, its imposing mix of late nineteenth- and early twentieth-century architecture, Germanic to the rooftops, blackened with coal-dust and beset by decay. Though it had escaped the ravages of the war, the Soviet advance passing it by to the north and the south, it had been less fortunate in the wake of the communist collapse. By the early 1990s, almost all mining operations in the region had been shut down, transforming it overnight from one of Poland's most prosperous areas to its poorest.

Men with nothing to do gathered on street corners and intersections and passed bottles of cheap vodka between each other, eyeing us suspiciously as we swept by.

On the narrow road leading out of Waldenburg to Ludwigsdorf, I asked Witkowski what had led him to believe that the Bell experiments had been an attempt by the Nazis to manipulate gravity.

There were scant clues, he admitted, but those that Sporrenberg had provided in his testimony seemed to add up to something. Many of the descriptions used by the Bell scientists did not gel with any of the accepted terms associated with nuclear physics, nor were there any obvious radioactive materials used in the experiments themselves. One of the terms Sporrenberg had picked up had been 'vortex compression'; another was 'magnetic fields separation'. These were physical principles that had come to be associated with the new wave of gravity and anti-gravity pioneers – people like Dr Evgeny Podkletnov, Witkowski said.

Podkletnov. Now, I began to pay more attention. If I'd interpreted Witkowski correctly, there was some kind of a relationship between the Russian's experiments with spinning superconductors – the effect he had tripped over in Finland when his assistant's pipe-smoke had hit that column of gravity-shielded air – and the effect produced by the Bell.

I needed to call Marckus and run some of this stuff past him; if, that is, he was still talking to me after my refusal to dig deeper in America.

Witkowski also claimed there were anomalies in the curriculum vitae of Professor Walther Gerlach that placed him firmly in the orbit of the gravity scientists, despite the fact that, ostensibly, his discipline was nuclear physics. In the 20s and 30s, Witkowski discovered, Gerlach had immersed himself in phenomena such as 'spin polarisation', 'spin resonance' and the properties of magnetic fields – areas that had little to do with the physics of the bomb, but much to do with the enigmatic properties of gravity.

A student of Gerlach's at Munich, O.C. Hilgenberg, published a paper in 1931 entitled 'About Gravitation, Vortices and Waves in Rotating Media' – putting him in the same ballpark as Podkletnov and the Bell. And yet, after the war, Gerlach, who died in 1979, apparently never returned to these subject matters, nor did he make any references to them; almost as if he had been forbidden to do so.

That, or something he had seen down the mine had scared him beyond all reason.

'The Germans ignored Einstein and developed an approach to gravity based on quantum theory,' Witkowski said. 'Don't forget that Einsteinian physics, relativity physics, with its big-picture view of the universe, represented Jewish science to the Nazis. Germany was where quantum mechanics was born. The Germans were looking at gravity from a different perspective to everyone else. Maybe it gave them answers to things the pro-relativity scientists hadn't even thought of.'

Kammler, Witkowski told me, had the ability to hoover up all scientific activity, whether it was theoretical or practical, through an SS-run organisation called the FEP, for *Forschungen, Entwicklungen und Patente* – researches, developments and patents. The FEP introduced the last major player in the story of the Bell and the Special Evacuation *Kommando*, an SS Obergruppenführer called Mazuw. A high-ranking general, Emil Mazuw had been able to acquire any significant technology, science theory or patent application that had come to the attention of the SS – and via its prolific security arms, there wasn't much that passed it by. The FEP, according to Witkowski's researches, operated independently of the *Reichsforschungsrat*, the Reich Research Council, but would unquestionably have had oversight of it.

My guess was that the FEP had been administered by Kammler's secret research cell within the Skoda Works. After the war, the Allied powers seized 340,000 German-held patents from the captured records of the Reich Research Council.

But here was something else: the cream of Germany's wartime scientific research that had been skimmed off by the SS and compartmentalised for its own use. Its value would have been immeasurable. Perhaps this really was what Kammler had returned to Czechoslovakia for in the death throes of the Reich?

There were other clues, too, but these, Witkowski repeated, were the physical traces at the mine itself and were best explained when we got there.

At Ludwigsdorf we followed the line of an old railway that had been built before the war to connect the region's network of industries to the outside world. As the road climbed, we found ourselves in a valley bordered on both sides by tall trees. A patchy mist clung to the upper branches as we ascended into the hills.

Then we rounded a bend and the scenery opened up. We were in a valley, one that had appeared out of nowhere. Witkowski parked on a patch of gravel overlooking some level ground. The railway track, overgrown from years of disuse, followed its median line, eventually disappearing from view behind a large derelict building whose tall arched windows rose cathedral-like into the mist. It was next to this building, Witkowski said, that the shaft of the mine disappeared below ground.

I stood on the edge of a deep cutting a short walk from the car. The valley was around 300 yards across, the trees either side of it so dense you'd never know the place existed unless you happened upon it.

The cutting, which was 20 feet high, was mirrored by an identical feature on the opposite side. During the war, this expanse of land had been an underground marshalling yard, Witkowski told me, wooden sleepers topped with turf hiding a six-lane section of track where the rail-head met the workings of the mine.

Even the buildings disguised the facility's true purpose.

High on the opposite bank was a large red-brick house, an original 19th-century feature of the Wenceslas Mine. But look a little closer, Witkowski said, and you could see where the Germans had tunnelled into the ground underneath and constructed a large concrete bunker, one of the many blockhouses scattered around the complex. All of the bunkers had been covered with earth and trees planted on top.

In short, the Germans had gone to a great deal of trouble to ensure that the place looked pretty much as it had always looked since mining operations began here at the turn of the last century, a clear indication that whatever had happened here during the war had been deeply secret. Now, half a century after the Germans had left, second-guessing the use to which it had been put, simply from the available physical evidence, was far from easy.

When, immediately after the war, the Poles repatriated the local German population, substituting German place-names for Polish ones in the process, they removed any witnesses who could have shed light on the SS' activities. Almost everything that was known about the Wenceslas mine had been handed down from Sporrenberg.

It had been run by the SS, had employed slave-labour and had been sealed from the outside world by a triple ring of check-points and heavily armed guards.

Beyond that, Witkowski said, we were into the realm of analysis, interpretation and detective work; the skills required for such a task being more akin to those of an archaeologist.

Although the locals had used the site for small-scale industrial activity in the years since the war – I could see freshly milled lumber stacked on a flat piece of land over by the valley entrance – it had been entirely neglected by the history books; its wartime use known only to the handful of people Witkowski had brought here.

That, I could understand. The place was as far off the beaten track as I'd ever been on the continent of Europe. I felt like I was tramping on ground that had hardly been touched since the Germans had pulled out more than half a century earlier.

The first thing to point out, Witkowski said, was the structure at the end of the valley. This, built before the war, was a power-station capable of burning a thousand tons of coal a day – enough to provide the valley with vast quantities of electricity. It had more than likely been this single feature that had drawn the SS to the site, Witkowski said, since it had helped to make the place entirely self-sufficient. After the war, when the Russians eventually made it into these hills, they would have found an abandoned complex given over to some quasi-military purpose, a mine shaft that had been flooded, possibly deliberately, and little more.

'With the help of Sporrenberg's testimony, we have an edge that was denied to the first Red Army units to enter this place,' Witkowski said. 'Come, I want to show you something; something very strange.'

We clambered back into the car and made our way down the track that led past the abandoned power-station. It was now dark enough for Witkowski to use his lights, the beams catching in the weeds that tumbled through the cracks in the pre-fab concrete roadway laid by the Germans during the war.

Whatever the SS had been moving around here, it had been pretty heavy-duty, Witkowski remarked.

The track cut through a wood, emerging in a clearing on the far side of the power-station. Caught in the headlights, rising out of the ground straight ahead, was a circular concrete construction 30 metres wide and 10 high. With its 12-metre-thick columns and horizontal beams, it was part-reminiscent of some ritual pagan edifice.

Witkowski parked the car, but left the headlights on, the beams bathing the columns of the object in bright white light.

'What is that thing?' I asked.

But Witkowski was already out of the door, crouching over something a few feet from the car. He was studying what looked like a partly exposed underground drain. Its concrete cover had cracked to reveal a duct about a foot across.

'This carried the electricity cable from the power-station,' he said. 'It disappears into the ground just beyond the car, but diverts via this thing.'

'What is it?' I asked again.

'I am not sure. But whatever it is – whatever it *was* – I believe the Germans managed to complete it. In this light it is difficult to see, but some of the original green paint remains. You do not camouflage something that is half-finished. It makes no sense.' He paused. 'There is something else. The ground within the structure has been excavated to a depth of a metre and lined with the same ceramic tiles that Sporrenberg describes in the chamber that contained the Bell. There are also high-strength steel hooks set into the tops of the columns. I think they were put there to support something; to *attach* to something. Something that must have exerted a lot of power.'

I looked at him. 'What are you saying?'

He took a moment before replying. 'I'm saying I think it's a test-rig. A test-rig for a vehicle or an engine of some kind. A very powerful one.'

That night I tried to call Marckus on my mobile, but the isolated farm house where we were staying was buried in a deep valley, every bit as remote as the one that housed the nearby Wenceslas Mine complex, and I couldn't get a signal.

The Polish couple who owned the smallholding let out a room to Witkowski whenever he came south from Warsaw to visit. They shared his curiosity about the mine and the neighbouring rat-run of uncompleted tunnels that were part of the SS-run 'Giant' underground weapons complex. For all they or anyone else knew, the tunnels extended under the ground I was standing on. But being only second-generation inhabitants of this region, their parents having been shipped in to replace the Germans who'd lived here for three hundred years previously, they knew as little about the history of the place as Witkowski and I did. This, undoubtedly, helped to give the place a detached, soulless air.

There was no anecdotal evidence to fall back on; the only evidence was history's imprint on the topography itself.

I needed Marckus. I needed to run every last detail of the mine past him to see if it rang any cherries in that weird analytical mind of his. I needed his guiding expertise again. Without some input from a physicist, the data on the Bell was nothing more than a bunch of science mumbo-jumbo.

As I collected my thoughts under the swath of clear, starlit sky between the ridgelines, I elected to concentrate on a matter I did know something about; something that had been nagging at me ever since Witkowski had mentioned it: the matter of the KG 200 unit at Opeln, the base that had operated the Ju 290 and 390 transport aircraft on detachment to the *Kommando*. It was a highly significant development.

The Ju 290 was quite a rare bird; the 390 even more so. The 290 had been a big four-engined aircraft designed for the maritime recon-naissance, transport and bombing roles. The first transports had been used in the Stalingrad airbridge in 1942, but towards the end of the war, the final variants were modified for extreme-long-range operations. Three such aircraft are known to have made flights to Japanese bases in Manchuria. The Ju 390 was a six-engined modification of the Ju 290 with even more impressive operating characteristics. Though only two prototypes were built, they clearly demonstrated the Ju 390's ability to mount ultra-long-range operations of up to 32 hours endurance. On one occasion, a Ju 390 flew to within 12 miles of New York and back again. It could also carry a very heavy payload.

Above all, though, the Luftwaffe had referred to these aircraft as 'trucks'.

A book in Witkowski's possession, *The Nuclear Axis* by Philip Henshall, had provided this one essential detail. Henshall had picked up on Tom Agoston's story of Himmler's request for a 'truck'. But Henshall had arrived at a markedly different interpretation of the data. When Kammler had telexed Himmler on 17 April from his Munich office – the last signal anyone ever received from him – refusing the Reichsführer-SS the use of a 'heavy truck' from the Junkers 'motor-pool', he hadn't been referring to a truck in the conventional sense. He had been referring to an aircraft. A long-range one with a heavy payload.

Kammler had been telling Himmler, his superior, that he couldn't have a Junkers 290 or 390 for his own use, because it was committed elsewhere. It could only have been for use by the Special Evacuation *Kommando*. Why else stamp such an ostensibly bland, innocent message *geheim* – secret? Why else would Himmler have been bothering himself with fucking trucks?

With Bormann, Hitler's deputy, in charge of the evacuation plan, Kammler would have been in a unique position to call the shots. Although Hitler viewed Himmler as a valued and trusted ally – certainly up until the moment his secret surrender negotiations were discovered – it was Bormann, Hitler's grey eminence, who had his ear.

Kammler could safely refuse Himmler permission to use a Ju 290 or 390 without fear of recrimination. With such aircraft at its disposal, the *Kommando* could have flown its cargo of documents, personnel and technology pretty much anywhere it wanted. Spain, South America – Argentina, even – would have represented no problem to such a long-range platform.

Jesus. It changed everything. What was the point of chasing Kammler, if he'd already shipped everything out?

After dinner, a good but simple meal of boiled eggs and local ham, I wandered back outside to take the air, leaving Witkowski inside to talk to our hosts.

I gazed at the ridgelines again. A few miles away, lay the Wenceslas Mine.

Part of me was transported back to the desert around Groom Lake, to the time when I'd stood vigil over the hills that shielded Area 51 from the outside world.

The part-completed Giant underground manufacturing complex and the Wenceslas Mine were a lot closer than I'd ever been able to get to the 27,000-feet-long paved runway on the desert lakebed. But the feeling they evoked in me was curiously similar.

The SS had installed a security system around the mine every bit as impenetrable as the cordon around Groom Lake. That, undoubtedly, was a part of it.

As was the fact that the technologies at the heart of both places were essentially a mystery to everyone but the people who'd worked on them.

It was not knowing what lay beyond those ridgelines that made me do this. It was the not knowing that drove me on.

All these sensations contributed to the almost tangible atmosphere that imbued both places.

I was consumed with the need to go on until the picture steadied.

I knew then that I was tired – more tired than I'd ever been. I had got to the point where my every living minute was filled with a need to know the truth of something this insoluble, and that it had got a hold of me so badly that my subconscious mind was trying to solve it in my sleep, failing in the process and turning in on itself, twisting raw thoughts and images into the worst kind of dreams. Here, in the depths of Kammler's kingdom, a thousand miles from the warm archives of the Imperial War Museum and the PRO, the scale of the crime had begun to cling to me like a second skin.

Amidst it all, though, a particular sequence of images had stuck in my mind since my visit to the mine. Try as I might to stop it, I couldn't. I saw it with my eyes open and I saw it with them shut. In the chill air between the ridgelines, I saw it now.

It started with the arrival of a convoy of Opel trucks a week before the end of the war.

The scientists would have had mixed feelings about leaving: a sense of regret at having to abandon the project; relief to be escaping the oppressive feeling of the place; joy at the knowledge they were getting out ahead of the Russians.

A late spring day, the guns a long way off, the scientists talking and smoking in threadbare suits, their bundled possessions to hand, as the convoy was readied for departure. Sixty-two scientists meant three trucks' worth; perhaps another five to transport the crates. Eight trucks in all, moving west, gears grinding on the steep roads through the mountains. A disciplined unit operating to a fixed schedule, the details drawn up by Sporrenberg's bureaucrats in Breslau.

Two kinds of troops attached to the unit: drivers and logistics personnel to oversee the evacuation procedure and an armed *sonder-kommando* group acting as escort.

To these special-action troops, the scientists were simply 'subjects'. The trick was to maintain their docility and compliance until the last possible moment.

After multiple 'operations' in the ghettos of Lodz, Warsaw, Minsk and the other occupied towns and cities of Central and Eastern Europe, the SS had found that by sticking to a pre-prepared book, the subjects could be kept as passive as cattle.

For some of the scientists, there may have been a tremor of foreboding as the three trucks peeled off from the main body of the convoy and headed into the forest.

But for most it would only have hit them when the tailgates dropped away exposing a line of troops with SS runes on their collars, their guns raised.

The Waffen-SS would have outnumbered the subjects two to one, their crushing shock and bewilderment combining with the SS' weight of numbers to make them respond without question to the barked orders as they moved towards the edge of the ditch. At no point, special-action orders decreed, should subjects feel that the executioners offered the remotest kind of hope.

And so, it had come to this. Two shifts, 31 bullets per shift, each shot three, maybe four seconds apart. The first victims face-down in the dirt for each member of the second shift to witness in a terrible moment of finality as he stepped up to the ditch, their grave, the proportions worked out in advance in some warm administrative office a long way from the front.

Sixty-two men meant a ditch ten metres long, two metres wide and one metre deep, the bodies stacked two high.

If the executioner tasked with placing his pistol at the base of each head observed the guidelines laid down for his benefit, the act of pulling the trigger was calculated to be no more arduous than switching off an electric light.

When the trucks left, the only thing to denote the scene of the crime would have been a mound of freshly dug earth half a metre high.

By the following spring, the mound would have collapsed to the exact same level as the ground around it. This, too, would have been carefully calculated according to the known decay-rates of human tissue and the ambient conditions of the ditch.

To preserve the secret of the mine, the SS had consulted not even a page, but a paragraph or two from the execution manual it had drawn up for the Holocaust. It had tapped on the crime of the century to do this to a handful of its own people.

Remember this, I told myself, should you ever be tempted to view this as a technology hunt in a remote place that has little or nothing to do with you.

Chapter 21

The next day, I was back into Germany, on my way to Austria, making good progress on the autobahns in the hire-car I'd picked up across the border, when the phone rang. It was Marckus. Hearing his voice, I felt a wave of relief, the cellular link between us suddenly feeling like a lifeline.

I spent the next 20 minutes briefing him on the mine. To begin with, he put up some resistance, trotting out his old belief that it was pointless pursuing the Germans for anything remotely as advanced as anti-gravity technology. But I had a feeling that Marckus was simply going through the motions.

The longer I talked, the more he listened. By the end, he was firing the questions at me.

I told him I had a hunch; that the Wenceslas Mine complex had been some kind of nuclear installation, the strange, henge-like construction being redolent of a reactor-housing, with a tile-lined coolant pool in its midst. I didn't buy Witkowski's test-rig thesis, but then again I wasn't dismissing it either.

As for the Bell, I had absolutely no idea what it could have been, but without my notes to hand, I couldn't give Marckus the kind of details he required to conduct an in-depth analysis.

I promised to send him an email with the facts that night, as soon as I got to my hotel in Bad Ischl, across the Austrian border.

'What happens in Bad Ischl?' Marckus asked.

I told him about the Junkers 'truck' business and how it had altered my plans. With long-range transport aircraft at his disposal, Kammler had had the means to ship the core secrets of the Skoda Works anywhere he wanted. And for all I knew, he had gone with them. It was pointless trying to pick up the clues of his disappearance in Ebensee or anywhere else for the time being. It was enough to know that the Nazis had, via Kammler's special projects group, a repository of technical secrets that had gone way beyond the V-weapons they had developed and used by the end of the war.

It had taught me other things, too. For years, I had read of the SS' involvement in exotic weapons technology, but dismissed it as inadmissible claptrap peddled by people with an unhealthy interest in its warped ritualistic ideology; this, despite Speer's solid documentation of Himmler's penchant for unconventional scientific solutions to Germany's dire military position in the closing years of the war. In *Infiltration*, Speer's account of how the SS eventually succeeded in establishing its own industrial empire within the visible economic framework of the Reich, he had listed some of Himmler's more absurd ideas: producing fuel from fir-tree roots, tapping the exhaust fumes of bakery chimneys for the manufacture of alcohol and producing oil in abundance from geraniums.

Speer also attests to Himmler's willingness to entertain any radical ideas for new weapons – especially those that leapt the current state of the art. One such was a proposal for turning the upper atmosphere into a giant high-voltage conductor, presumably for the purpose of frying the Allies' B-17, B-24 and Lancaster bombers as they flew into Reich airspace. These ideas, many of which were based on unsound science, came to naught, but it would only have taken one far-reaching proposal underpinned by some solid scientific reasoning for this vast array of funded research activity to deliver the pay-off so desperately sought by the Nazi leadership.

Like lasers, for example. Or an atomic bomb.

It had also taught me just how effective – and this word, of course, glosses over the grotesque cruelty of its methods – the SS security machine had been.

The *Kammlerstab* had been protected by a triple ring of counter-intelligence agents to prevent word of its activities from leaking. If it hadn't been for Voss and Agoston, the existence of the special projects group at Skoda might never have surfaced at all.

It had been the same story at the Wenceslas Mine. A combination of active security measures and a neighbouring populace that had been rounded up and removed *en masse*, and its wartime activities had all but disappeared from existence.

But there was an even more powerful reason why none of this had ever properly surfaced before.

By involving the concentration camps, the SS had unwittingly set the seal on any serious postwar investigation of the science and technology it had pursued during the conflict. Because it had been the SS, not so much Speer's Armaments Ministry, that had backed so-called high-payoff, 'visionary' projects with funding, German industry found itself in league up to its collective neck with the perpetrators of the Holocaust.

Many of the companies that had benefited from SS money, and employed concentration camp labour in the process, are still in existence today.

Bosch, Siemens, Zeiss and AEG all maintained highly secretive research and development operations in Lower Silesia.

It remains in their interests for their activities at this time to stay undocumented. I doubted, in any case, even if they were to open up their archives, whether anything of real value remained in them. Like the records pertaining to Kammler's past, I suspected they'd long since been cleaned out.

This, I said to Marckus, was how secrets came to be locked away, buried for all time, leaving nothing but myths and legends in their aftermath.

Maybe the Nazis had initiated a flying saucer programme; maybe some of the technology had borne fruit. There was certainly enough anecdotal evidence to support the view that a variety of disc-shaped craft had flown before the end of the war. It was simply that there was no *proof*.

My journey into Kammler's kingdom had opened my mind to possibilities I wouldn't have begun to entertain a few months earlier. And in the process of this willingness at least to confront new ideas, I had found myself drawn to Bad Ischl.

It was there, before leaving London, that I had arranged to meet with the family of Viktor Schauberger.

I had picked up on the Schauberger story via the Legend forwarded to me by my friend and former colleague, Lawrence Cross, in Australia. I had dismissed it out of hand because not being an engineer but a simple forester, what this Austrian inventor was said to have achieved technically hadn't made any sense. And the phone conversation I had had with his son Walter, who back in 1991 had urged me to visit the family-run 'bio-technical institute' in the Salzkammergut mountains so I could make up my own mind, had only served to alienate me further.

But Bad Ischl was less than 15 minutes' drive-time from Ebensee. And a lot had changed in the interim. I knew, for example, thanks to Podkletnov and his lab assistant's pipe-smoke, that whatever anti-gravity was or wasn't, it had to be induced by something highly unconventional. And Schauberger's approach – if the Legend as relayed to me by Cross contained even a grain of truth – was certainly unconventional.

The Schaubergers maintained a large archive dedicated to Viktor's work in their house, the location of the institute, on the outskirts of Bad Ischl. If they would allow me to take a look at it, perhaps I would be able to find something, some proof, that the Germans really had developed a new and exotic propulsion medium.

They were reluctant to do so and I could understand why.

It was well documented in the legends that had grown up around him that Viktor Schauberger had worked for the SS.

In my initial assessment of Schauberger's work, this association seemed to have been drawn straight from the mythology surrounding SS activities during the war and had merely compounded the case against him.

Now, of course, it was quite the reverse.

I'd reached the end of the road; told Marckus everything I'd learned. It was only when I finished that I realised he hadn't interjected once.

I thanked him for listening. He stopped me just before I cut the link.

'Call me, won't you, if you need to talk. I mean it. Day or night. Sometimes it can be good to download, clear the mind for what lies ahead.'

As I journeyed on towards the mountains, I felt better about Marckus than I had in months. It was difficult to put my finger on, but it had something to do with the fact that, for once, I was telling him about material of which he had no prior knowledge. I wasn't entirely sure why, but for the first time I felt I could trust him.

Something about Marckus' final tone of voice told me something else. I felt sure that he knew exactly what I was going through.

As soon as I crossed the Austrian border, I took a break in a lay-by, got out the laptop and pulled up Cross' notes on the Legend. The relevant portion, the part devoted to Viktor Schauberger, boiled down to the following precis:

Late in the war, despite being close to pensionable age, Schauberger was called up for active duty in the German Army. Soon afterwards, he received orders to report to an SS institution in Vienna. From there, he was taken to the nearby concentration camp of Mauthausen, informed by the camp commandant that his inventions had received the blessing of Reichsführer Himmler himself, and ordered to hand-pick a group of engineers from amongst the prisoners. This would be the 'team' that would help him complete his work on an energy device of radical design that Schauberger had begun working on before the war. If he did not comply with this order, the commandant informed him, he would be hanged and reprisals instituted against his family. Schauberger did as he was told.

By 1944, serious work had begun on a Schauberger machine that appeared to have a dual purpose: first as an energy-generator and second as a powerplant for an aerospace vehicle of saucer-like appearance. Descriptions of the workings of this generator, sometimes referred to as a 'trout turbine', were always woefully inadequate. One such relayed the apparent fact that 'if water or air is rotated into a twisting form of oscillation known as "colloidal", a build-up of energy results, which, with immense power, can cause levitation.

'Details of the chronology are hazy,' Cross went on, 'but it seems that Schauberger's team achieved success just a few days before Germany surrendered. One of the scientists who'd worked with Schauberger reportedly said that at the first attempt to run the machine "the flying saucer rose unexpectedly to the ceiling and then was wrecked. The apparatus functioned at the first attempt . . . and rose upwards trailing a blue-green and then a silver-coloured glow."

'The destroyed craft had a diameter of 1.5 m, weighed 135 kg, and was started by a small electric motor with take-off energy supplied by the trout-turbine.

'The next bit is really weird,' Cross wrote. 'A few days later an American group reportedly appeared, who *seemed to understand what was happening and seized everything* [his emphasis, not mine]. Schauberger was kept under "protective US custody" for six months to a year, no one seems sure quite how long. It appears some of his work was classed as "atomic energy research" because this was the area the US forbade him to continue with. I just don't know what happened to him afterwards.'

I kept all of this in mind as I drove through Ebensee, scanning the mountains for a glimpse of the entrance to Kammler's tunnel-system: the Zement complex that would have developed and manufactured the A-9/A-10 intercontinental rocket. I saw nothing but for some roadsigns directing visitors to a Holocaust Memorial. Zement had become a tourist attraction.

Following directions from Schauberger's grandson, Joerg, I pushed through Bad Ischl, following the River Traun through a patchy expanse of woodland until, on the left-hand side of the road, I spotted the institute, a large imposing house with a tower rising from one of its wings, the initials 'PKS' painted in letters six feet high on one of its walls. It was here that the Schauberger family organised periodic lectures and seminars devoted to the heretical scientific principles promulgated by Viktor and his son Walter. Walter had since died, but *his* son Joerg continued to push the ideas that his father and grandfather had promoted for the creation of a better world.

All I knew about the Schaubergers was what I had read; that Viktor, through his early work as a forester, had developed a passionate set of theories about the essential characteristics of air and water as living, energising media and that he viewed Nature as a complex interaction of forces that constantly created or re-invigorated matter – against the orthodox view that matter is in a natural state of decay.

Walter, a trained mathematician, had taken his father's theories a step further, merging Viktor's views with Pythagorean concepts of harmonics and the laws of planetary motion developed by Johannes Kepler. It was no coincidence, Walter believed, that a carefully cut cross-section through a Pythagorean hyperbola or the elliptical planetary orbits plotted by Kepler happened to be egg-shaped, a form that was one of Nature's highest expressions of energy and strength. PKS, I remembered now, stood for Pythagoras-Kepler System.

It sounded more than a little like New Age craziness, the last place that someone with my background and training should have gone out of his way to visit; but then again, I reminded myself, maybe it was actually no crazier than NASA's Garry Lyles telling me that man would be journeying to the stars by the end of the 21st century in a spacecraft that used some kind of warp-drive or worm-hole.

I eased the car onto the Schaubergers' property. The institute lay at the end of a long drive. The snow-capped tops of the mountains beyond were dazzlingly bright against the blue sky.

Joerg Schauberger, Viktor's grandson, greeted me on the steps of the house and led me to the basement area where a series of small-scale experiments had been set up to demonstrate the theories developed by his father and grandfather. Viktor's maxim had been 'comprehend and copy Nature' and central to this thesis had been his understanding of the vortices that occurred naturally in the environment.

These 'energy spirals', as he saw them, creative whirlpools of Nature, were visible everywhere: from the spiral galaxies of the outer cosmos to the power of a tornado or simply in the whorled horn of a kudu's antler or the double-helix of a DNA strand.

Nature, Viktor Schauberger had believed, employed the vortex as its most efficient conduit for the transmission of energy.

When this three-dimensional spiralling energy pattern was channelled inwards, not outwards, in a process Viktor called 'implosion', it became endowed with 'higher order' properties – characteristics which Viktor himself described as 'atomic' (although his understanding of the word was quite different from that of a nuclear physicist's).These properties were capable of generating phenomenal levels of force. Transposed into machinery, Viktor had even coined a term for it: 'bio-technology'.

One such device, an implosion-based generator, stood disused and rusting in the corner of the basement area where Joerg Schauberger and I now viewed the fruits of his forebears' lifework.

Joerg Schauberger had signalled during our phone conversations his reluctance to prise open the family archives to a complete stranger. But he was content for me to come to Bad Ischl to plead my case. I was glad that I had. He was roughly my age and had an open, honest face. I liked him immediately. I could see that he was weighing me up every bit as carefully as I was assessing him. Both of us were entering uncertain territory. For me, his grandfather's experiments bordered on the incredible. But Joerg wasn't interested in publicity. This in itself gave me confidence. Whatever lay in the archives, I knew it would be fresh, untainted evidence, free from any kind of interpretation or bias.

Joerg was proud of his family's achievements. In the lofty, baronial entranceway of the family institute, he had shown me a cutaway section of a log flume that his grandfather had built in the forests for the efficient removal of lumbered trees.

It was then that I realised the impression painted of Viktor Schauberger in the Legend was a misleading one. Schauberger wasn't a forester in the strict sense of the word. He was an engineer; and from the look of the machinery around the house an extremely accomplished one. His milieu happened to be the forest.

Now, as we stood beside the implosion machine, his grandson spelled out his fears. Others had documented Viktor Schauberger's experiments, he told me; there was little, he believed, I or anybody else could add to this body of knowledge – the more so in my case, as, by my own admission, the science of the processes involved wasn't principally what had led me here.

I was after proof of a technology that conventional science said was impossible.

If, as the Legend had it, his grandfather had cooperated with the SS, what did his family and its reputation possibly have to gain from my seeing the archives?

I had thought long and hard about this during my drive through Germany, but whichever way I cut it, there was nothing reassuring to say.

So I told him instead about the journey that had brought me to his house; about the series of statements in 1956 by US aerospace companies that the conquest of gravity was imminent, that all that was required to make it happen – to usher in an era of clean, fuelless propulsion technology, of free-energy – was money and a little application from the US government.

I told him about Thomas Townsend Brown's efforts at much the same time to interest the US Air Force and Navy in an electrogravitically powered Mach 3 flying saucer and how aspects of Brown's ideas had emerged in the B-2 Stealth Bomber more than three decades later.

I told him how Evgeny Podkletnov had defied the theoreticians of NASA's Breakthrough Propulsion Physics programme by engineering a device that shielded gravity by as much as five per cent and how, officially, Brown's work had been discredited by the US military – despite its very real application to the B-2.

We spoke, as well, about the glaring discrepancy between General Twining's view in 1947 that the development of a manned aircraft with the operating characteristics of a UFO was 'within the present US knowledge' and the fact that, even in the black world of modern US aerospace technology, there were so few traces of this knowledge it was as if it had never existed.

Which had brought me to the Germans.

There was, via the Kammler trail, a mounting body of evidence that the Nazis, in their desperation to win the war, had been experimenting with a form of science the rest of the world had never remotely considered. And that somewhere in this cauldron of ideas, a new technology had been born; one that was so far ahead of its time it had been suppressed for more than half a century.

Joerg Schauberger's grandfather, I felt sure, was integral to the truth of what had happened at that time. But unlike the Wenceslas Mine, where few traces of Sporrenberg's court testimony remained on view, Viktor Schauberger had left records. That, I told his grandson, was why I was here.

At this, he gave an almost imperceptible nod and indicated we should head upstairs. It was getting late and there was a lot of material we needed to get through in the short space of time I was in Bad Ischl.

Like me, he said, pausing at the base of the steps that led to the upper levels of the house, he was only interested in the truth. In the complex world his grandfather had lived in, it was a commodity that was not always so easy to get a handle on.

It was then that I understood. Joerg Schauberger had been trying to assemble his own jigsaw – one that explained the intricate mystery of his grandfather's life.

I held pieces of the puzzle he didn't possess.

In regulating the movement of water via dams and irrigation schemes, man had obstructed the complex, vortex flow patterns that water naturally adopted for itself, Viktor Schauberger said, causing widespread degeneration and decay in the ecosystem. Harnessed in this unnatural state, water, he came to believe, acted as a pollutant upon the earth, not a life-force.

In seeking to come up with methods to restore the earth's ecological balance, he turned to the development of 'bio-technical' machinery and promoted his ideas through books, some of which came to the attention of Berlin.

In 1934, a year after Hitler came to power, Schauberger was summoned to the German capital to explain to his Führer and fellow-countryman how processes of natural motion and temperature and the vital relationship between soil, water and vegetation combined to create a sustainable and viable society; a society, in effect, that was at ease with itself.

At the meeting was Max Planck, the great German physicist and pioneer of quantum theory, who when asked by Hitler at the end of Schauberger's talk what he thought of his theories replied testily, 'Science has nothing to do with Nature' and withdrew from the discussion. Hitler nonetheless asked his technical and economic advisers what could be done to incorporate Schauberger's ideas into the country's four-year economic plan, but nothing ever came of it.

For his part, Schauberger told Hitler that the short-termism of Germany's economic strategy would undermine and ultimately destroy the country's biological foundations.

As a result, he said, the thousand-year Reich would be lucky if it lasted ten.

By using an 'impeller' – a propeller that induced an inward, instead of outward-flowing motion – to draw water through a tube, a flow-pattern Schauberger referred to as 'centripetal [the opposite of centrifugal] force', he found that the output he was getting was nine times greater than could be achieved with a conventional pressure turbine.

His early egg-shaped implosion machines were also generating extremely high vacuum effects.

By substituting air for water, Schauberger began to envisage a device that with some refinement could be put to use as a radical form of aero-engine; one that sucked rather than pushed its way through the atmosphere.

In 1939, he conceived of a device that could be put to use either as an energy generator or as a powerplant for aircraft or submarines. In the application he lodged to the Reich Patent Office the following year, Schauberger described the essential characteristics of this machine as a 'multi-stage centrifuge with concentrically juxtaposed pressure chambers'.

Shortly afterward, he wrote to his cousin that he had invented an aircraft that didn't make any noise.

With these devices, Schauberger realised he had created an entirely new methodology for propelling vehicles through air and water.

As Joerg Schauberger and I settled into the archive, which was crammed full with the old man's files and papers, it was clear that Viktor Schauberger had documented every turn of his career in meticulous detail.

Through his letters, duplicates of which he always placed on file, it was possible to paint an intricate profile of the man and his inventions. The challenge, even for his family, lay in decoding the shorthand Schauberger had used to decribe his work.

During the war, he had concentrated on the development of several types of machine – the Repulsator and Repulsine for water purification and distillation, the Implosion Motor for electricity generation, the Trout Turbine for submarine propulsion and a 'flying saucer' that used air instead of water as its driving medium. Because they all worked on the same principles, Schauberger tended to interchange the names of these devices and their applications when it suited him.

So, when Viktor wrote in 1940 that he had commissioned a company in Berlin called Kaempfer to build a 'Repulsator' it wasn't immediately apparent what this machine was for. It was only when he ran into contractual difficulties with Kaempfer, which was having enormous problems manufacturing the machine to Viktor's demanding speci-fications, that its function became clear.

By February 1941, Viktor had switched contractors to the Kertl company in Vienna, and here, in correspondence, he described the prototype (which he was building at his own expense) as having a two-fold purpose: to investigate 'free energy production' and to validate his theories of 'levitational flight'.

The machine relied on a turbine plate of waviform construction that fitted onto a similarly moulded base-plate. The gap between the plates was whorl-shaped, mimicking the corkscrew action of a kudu's antler.

Having drawn air in via the intake, the rapidly rotating turbine propelled it to the rim of the rotating mass under centrifugal force.

The vortex movement of air created by the waviform gap between the plates led to its rapid cooling and 'densation', producing a massive reduction in volume and generating a vacuum of enormous pressure, which sucked more air into the turbine.

The machine required a small starter-motor to commence the process

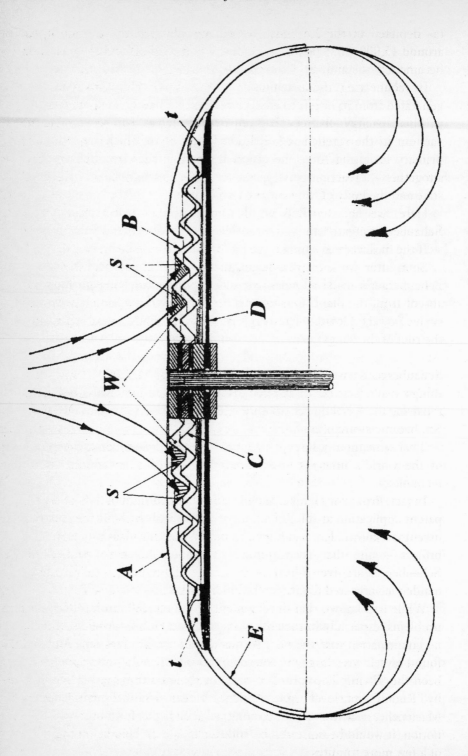

(as depicted in the Legend), but having whipped the turbine up to around 15,000–20,000 rpm the motor was turned off and the operation became self-sustaining.

By connecting the machine to a gear shaft, electricity could be generated from it; or left to its own devices, it could be made to take off.

This capacity to fly Schauberger partly attributed to the creation of the vacuum in the rarefied region immediately above the plates. But the primary levitating force, he claimed, was due to some other process altogether – a reaction between the air molecules in their newly excited state and the body of the machine itself.

Here, we had touched on the heart of the matter: had Viktor Schauberger created an 'anti-gravity' device?

If the files were anything to go by, the answer was unequivocal.

Soon after work on the device re-started at Kertl, an associate of Schauberger's made an unauthorised test-run while Schauberger was absent from the plant. During this experiment, the machine generated such a powerful levitational force that it shot upwards, smashing against the roof of the hangar.

Schauberger's correspondence makes it clear that March 1941 was when things really started to happen. It was then, while work began on repairing the Kertl device, that the Gestapo, the secret police arm of the SS, became aware of his work.

That same month, he reported that Professor Ernst Heinkel, inventor of the world's first jet aircraft, was also showing an interest in his technology.

In fact, Professor Heinkel had illegally obtained sight of Schauberger's patent application at the Reich Patent Office and, without the Austrian inventor's knowledge, had begun to incorporate his ideas into a Heinkel project – one that is presumed by some who have studied the Schauberger archive to have been the He 280 fighter, which made its maiden jet-powered flight on 30 March 1941.

While it is known that development of the He 280 ran into technical problems, especially in the area of its troubled HeS 8 turbojets, there is no confirmation that the He 280 was the aircraft in question. And so I found myself wrestling with a heretical notion. It could just as easily have been the 'Flying Top' proof-of-concept vehicle supposedly being built by Rudolf Schriever in a 'garage' under Heinkel's guidance at Marienehe, near Rostock on the Baltic coast. Try as I might to resist the notion, it would be bolstered by additional data on Heinkel in the space of a few more minutes.

The Schriever story was part of the legend that had grown up around Rudolf Lusar's *German Secret Weapons of World War 2*, the book that had drawn me into the contentious area of German flying disc research. In it, Lusar, who had been well informed on all other aspects of Geman secret weapons activity, made the bold assertion that the first Nazi flying saucer projects commenced in 1941 and that the research was centred on two places: Prague and Breslau.

Heinkel coerced the patent office into restricting the use of Schauberger's technology to water purification and distillation applications, leaving Heinkel free to make fraudulent use of Schauberger's innovations in his aircraft projects. When the Reich Patent Office offered Schauberger a patent on this basis the Austrian refused and consequently had his application turned down.

In May 1941, Schauberger was formally approached by the 'Gestapo', who served him notice (in the company of his patent lawyers) that they would not restrict his work – far from it – but in future he had to pursue it in secret.

This meeting appears to be the trigger for a period of highly covert activity in Schauberger's life, about which he leaves few clues in his letters. Weeks after his meeting with the Gestapo (which, in Schauberger-speak, could just as easily have been any SS personnel), he wrote to his son Walter saying he was in Gablonz (present-day Jablonec) in Czechoslovakia and that 'what I am doing is secret.'

Today, Jablonec is almost exactly where the Czech, German and Polish borders meet. During the war, it was bang in the heart of Kammler's high-tech kingdom.

In a letter the following month, Schauberger told his son that he had an agreement with a 'factory' in the Sudeten region which was undertaking unspecified 'research'. This factory appears to be one and the same place Schauberger had mentioned earlier: Gablonz.

Looking at the map and trying not to be seduced by the Legend, I noted that Gablonz was conveniently located between Prague and Breslau.

In the same correspondence, Schauberger reveals two other facts of significance: first that work at Kertl (apparently on repairs to the device that smashed against the roof of the hangar) is 'going so slowly it is as if someone is deliberately trying to slow it down'; and that he has become aware that Heinkel has stolen his ideas.

Schauberger also reveals that his source in the Heinkel matter is a member of the 'secret police' – a term Schauberger used interchangeably with 'SS'.

In July 1941, Schauberger describes how 'an intermediary' has approached him in a bid to rectify certain problems Heinkel was experiencing with the technology he had lifted from Schauberger's patent application. With ill-concealed *schadenfreude*, Schauberger reveals that Heinkel's engineers have made a mistake copying his design, and that he knows what the error is. Through his dialogue with the intermediary, Schauberger believes that Heinkel is trying to substitute conventional engine technology with technology of his own.

On the face of it, this ought to mean one of two things: that the compressor technology of the troubled He 280's HeS 8 turbojets had been marked for replacement by Schauberger turbine components, the 'centripetal compressors' of his successful 1936 patent application; or that Heinkel was attempting to swap complete Schauberger powerplant units for the underperforming HeS 8 turbojets.

The first explanation makes little sense. Adding a 'centripetal compressor' – one that causes air to flow radially inwards – would have destroyed the carefully tailored flow dynamics of the HeS 8 engine, which was a centrifugal gas turbine based on a principle diametrically opposed to the Schauberger implosion technique.

The second explanation is equally vexing. Strapping Schauberger's engines, which worked on a suction principle, onto the He 280 would have made a total nonsense of the aircraft's graceful aerodynamics.

History, in any case, tells us that the Air Ministry (RLM) would only give the He 280 a production order if Heinkel substituted the HeS 8 engine for BMW 003 or Jumo 004 axial-flow turbojets. Neither of these would fit, so the entire project was scrapped.

There has never been any mention in Heinkel files, so far as I am aware, of any association between the He 280 project and Viktor Schauberger.

This process of elimination allowed me to entertain the notion with greater confidence than I had earlier that Heinkel had been working on some altogether different form of air vehicle – one that would have worked best with a Schauberger engine, not a gas turbine; and one that has never come to light.

It is interesting to note that the legend of the Schriever Flying Top makes specific mention of the fact that construction of full-sized disc-shaped vehicles was transferred from Heinkel's Marienehe facility at Rostock to Czechoslovakia in 1943.

From late 1941, Schauberger's correspondence makes it plain that he is busily engaged in secret work in Czechoslovakia. In late 1941, he found himself in a 'weapons factory' at Neudek (now Nejdek) near Karlsbad

(Karlovy Vary), around 120 km southwest of Gablonz. In December 1941, he was back in Gablonz eagerly awaiting the arrival of a starter-motor – presumably, the little electric engine that was needed to kick-start his implosion device. At one point he tells his son that he is having to make frequent visits to the train station at Reichenberg (today, Liberec in the Czech Republic) to see if the 'apparatus' has arrived.

In early 1942, Schauberger transferred to the Messerschmitt prototype works at Augsburg in Bavaria. There, he worked on the further development of a Repulsator implosion machine. In his correspondence, it is not clear if this is the same machine as the ones on which he was working in secret at the two locations in the Czech Republic, although it seems unlikely. What appears to be the case is that Schauberger's work was central to a number of parallel activities during 1941 and 1942: Heinkel's mysterious work at Marienehe, the two secret weapons factories in Czechoslovakia, the Kertl company in Vienna and Messer-schmitt at Augsburg.

From Schauberger's writings, we know what happened to the Kertl and Messerschmitt machines. Work on the former ceased due to the company's inability to obtain essential supplies for the repair of the damaged device. This is not unreasonable, given Germany's growing inability to get hold of strategic raw materials from this period on, but Schauberger still voices his doubts, being convinced that somebody has ordered the work to be slowed deliberately.

The Messerschmitt device suffered a catastrophic failure due to the use of substandard casting techniques (and probably low-grade metal alloys) during the manufacture of the turbine. According to Schauberger it suffered a 'meltdown' as soon as it reached its maximum rotational velocity. A company in Vienna, Ernst Kubiznak, was ordered to repair the device under a 'Führer directive', but this, like the Kertl work, appears to have come to naught.

This leaves the Heinkel work and the secret activities in Czechoslovakia.

After the war, when Schauberger picked up on rumours of German flying saucers that supposedly flew near Prague during the closing stages of the conflict, he voiced his conviction that these were developments of his own ideas – and, given the way Heinkel had behaved, this may have been the case. It was already quite clear that the Sudeten region was the epicentre of some of the Nazis' most secret and unconventional research of the war period. But flying saucers? Anti-gravity propulsion?

In April 1944, after a year-long period of inactivity, Schauberger was summoned before a 'draft panel' of the Waffen-SS.

In June, he travelled to Breslau, ostensibly to join an SS–Panzergrenadier division. But more than six months before the Red Army had battled its way to the gates of the city – before Gauleiter Hanke had sifted the citizenry of Breslau for men far fitter than Schauberger – it is ludicrous to suppose that the SS would have recruited an Austrian of his age to fight in the defence of a patch of the Fatherland hundreds of kilometres from his home. In April 1944, Schauberger was 59 years old.

Given the SS penchant for extreme secrecy, coupled to the fact that Schauberger had already been up to his eyes in secret programme work in Czechoslovakia, it is infinitely more likely that the Panzergrenadier story was a cover and that Schauberger was taken to Breslau for a specific purpose; one that played to his primary talent as an inventor.

This would have been an ill-disciplined leap of faith, a lapse in cold reasoning on my part, but for one thing.

By June 1944, the SS was almost a year into its take-over of the German secret weapons industry. Bit by bit, the Luftwaffe and the Wehrmacht were being eased out of arms production. In spite of the failures at Kertl and Messerschmitt, there is ample justification – based on what happened next – to assume that the Czech-based projects had validated at least some of Schauberger's ideas.

A month previously, a month *before* Schauberger went to Breslau, he was ordered to Mauthausen concentration camp to hand-pick a team from among the inmates to start building as many as five different types of Schauberger machine. This is confirmed by the Schauberger archive. The SS wanted him to stop 'tinkering around with prototypes and begin serious construction work'.

Somewhere along the line, someone must have been convinced that the technology worked.

The essential point, the point that made the hairs on the back of my neck stand on end as I settled into the part of Schauberger's diary that dealt with the last six months of the war, was one of which Viktor himself would have been unaware as he arrived outside the stone walls of Mauthausen camp in that summer of 1944. The big picture, which was in the possession of a mere handful of individuals in Berlin, was that the concentration camps were the designated *production* engines of the 'state within the state', as Speer had called the SS; the grim new powerhouses of the Reich economy.

Viewed this way, it shed light on Schauberger's frantic activities during 1941, when his feet had hardly touched the ground as he dashed between project locations in Germany, Austria and Czechoslovakia, a time that could now be re-interpreted as the start-date for prototype work of a diverse and highly secret nature; work that continued until early 1944, when the SS became satisfied that the machines were ready for manufacture.

The immediate task of Schauberger's hand-picked team at Mauthausen was to produce engineering drawings from which production work could begin.

Schauberger's diaries tell us what these machines were. One was a water purifier. Another was an energy device capable of generating high-voltage electricity. A third was for 'biosynthesising' hydrogen fuel from water. A fourth was a machine that 'naturally' produced intense heat or cold. The fifth, dubbed the 'flying saucer', was the unconventional aero-engine that had come to the attention of Heinkel and others during 1941. It was this latter device that I now focused on.

Schauberger set up design offices in the SS technical engineer school, in the Rosenhugel district of Vienna.

Working under oppressive conditions, effectively at gunpoint, he managed to gain a number of key concessions from the SS. Telling them that he needed personally to select his team from Mauthausen's inmates and that these men should be removed from the camp environment, fed and clothed properly if they were to work productively, he was able to extract around 25 former technicians from the camp and set them up in workshops nearby.

The first set of pre-production drawings were signed off in August 1944 and delivered from the engineer school to these facilities. Schauberger moved between the two sites, his visits to Mauthausen increasing in frequency as metal was cut on the propulsion device, which he referred to sporadically in the text as the Repulsine (even though this name had come to be associated in former years with other projects).

By early October 1944, Schauberger's diary shows that he was already starting to mill components for the Repulsine. On 6 October, he made the first pre-production drawings of a part of the device he refers to as the *Schaufelentwurf*, a design for a cone-shaped air-intake that bolted onto the top of the disc.

Five days later, work is interrupted by the first of many air raid alerts as Vienna and its environs are hit by Flying Fortresses on daylight missions against local industrial targets. Mauthausen was close to Linz, home of the Hermann Goering Works, a giant heavy engineering complex. It is clear from these and other entries that Schauberger was racing against the clock to complete his task.

On 14 October, he wrote: 'I wish I could work longer, but I have to stop at four in the afternoon. There is a shortage of materials and tools. I had to construct parts today from an old tank.' Because much of the raw material was of low quality, he instructed his team to be particularly mindful of heat stresses and welding quality.

It was also clear that he was being forced to shift between projects as demands required. On 23 October, he turned his attention to developing a more efficient carburettor for Opel Blitz army trucks, presumably on the instructions of the SS, an activity that distracted him for the best part of ten days. On 4 November, one of the water-based devices needed his design input. On 26 November, it was the turn of the Repulsine again – the electric starter motor needed to be checked and fitted.

The team worked on the Repulsine on Christmas Day and on into the New Year, all the while having to contend with power cuts and air-raids. Finally, on 28 February, Schauberger moved his operations centre to the little village of Leonstein in Upper Austria. The decision was a fateful one on two counts. It not only took him away from the bombing, but placed him squarely in what would later become the US zone of occupation. Had he stayed in Vienna, Schauberger would have found himself in the Soviet zone of influence.

In light of what happened to him in the days, weeks and months that followed the end of the war, it might have been the softer option.

On 5 April 1945, the diary reveals that final assembly of the Repulsine had commenced. A month later, it was ready to go. But the end, when it came, was almost surreal. Instead of firing up the disc for its first test-run as planned on 6 May, Schauberger awoke to find that the SS officers charged with oversight of the operation had fled into the night. The team stopped work on 8 May, hours before the surrender of German forces took effect at a minute past midnight on 9 May.

The SS machine had not flown in the last days of the war, as the Legend had maintained, but in almost all other respects it had held up remarkably well.

The diary had made it clear that Viktor Schauberger had built a machine that had flown earlier in the war at Kertl (and almost certainly during Schauberger's secret period of research in Czechoslovakia). It was also quite clear that the device's modus operandi was wholly unconventional – that is to say, the method by which it generated lift was insufficiently explained by current scientific knowledge.

The diary had given me something I could believe in at long last.

Via Schauberger, the Nazis had been deeply involved – no question – in what can only be described as flying saucer technology. And from this flowed the corollary that other parts of the Legend were perhaps also based on fact.

Couple the Schauberger evidence with what I had seen in the *Lusty* files and suddenly it wasn't so hard anymore to believe that the Germans had developed prototype remote-controlled air vehicles – craft that had the ability to latch onto the exhaust wakes of Allied bombers or the infrared signatures of the aircraft themselves – that explained many of the foo-fighter sightings of 1944 and 1945.

It was also highly likely that some of these 'conventionally powered' German disc-shaped craft were test-beds for Schauberger-type propulsion systems.

And therein lay another subset to the mystery.

Something about this whole strand of development had conspired to make it the most classified form of technology in existence. Even more so than the bomb.

Unlike the bomb, however, this was a secret that had held for more than 50 years.

Days after the end of the war, US intelligence agents found Schauberger in Leonstein and apprehended him. Exactly as the Legend had it, the agents, who were almost certainly Counter-Intelligence Corps – the same outfit that had detained and interrogated Skoda's director, Wilhelm Voss – were remarkably well informed about his entire operation. It was as if, Schauberger noted later, someone had guided them directly to him.

Across the country, at almost exactly the same moment, Russian intelligence agents located and entered his apartment in Vienna. They removed anything of value they could lay their hands on, then blew the place up, presumably to prevent anyone else from discovering information they had overlooked.

For the next nine months, until March 1946, Schauberger was held under close house arrest and extensively debriefed by US technical intelligence about his activities during the war. Years later, he disclosed that it was his knowledge of 'atomic' energy that had got the Americans so vexed. This is understandable. The merest suggestion that Schauberger had engaged in nuclear research would have set alarm bells ringing at the local headquarters of Boris Pash's Alsos agents, the US team charged with scouring the former Reich for atomic weapons activity.

And yet, it would have taken the CIC and Alsos days, if that, to ascertain that what Schauberger had really been up to had nothing to do with a Nazi bomb or even atomic energy production – at least, not in the sense Los Alamos understood it.

That night, back at my hotel, I applied myself to a study of the

Schauberger effect that had been written up by a researcher called Callum Coats. I had picked up a copy of one of his books at the Schauberger institute. In it, there was a description of what happened when a Repulsine was rotated at 20,000 rpm. The high rotation speeds appeared to cause the air molecules passing through the turbine to pack so tightly together that their molecular and nuclear binding energies were affected in a way that triggered the anti-gravity effect. 'A point is reached where a large number of electrons and protons with opposite charges and directions of spin are forced into collision and annihilate with one another,' Coats wrote. 'As lower rather than higher orders of energy and the basic building blocks of atoms, they are upwardly extruded as it were out of the physical and into virtual states.'

Virtual states? What the hell did that mean?

A few paragraphs on, Coats elaborated on this theme: 'Through the interaction between centrifugal and centripetal forces functioning on a common axis, he was able implosively to return or re-transmute the physical form (water or air) into its primary energetic matrix – a non-spatial, 4th, or 5th dimensional state, which has nothing to do with the three dimensions of physical existence.'

'I stand face to face with the apparent "void", the compression of dematerialisation that we are wont to call a "vacuum",' Schauberger had written in his diary on 14 August 1936. 'I can now see that we are able to create anything we wish for ourselves out of this "nothing".'

More than 60 years after Schauberger had consigned these thoughts to paper, I had stood in the Austin, Texas, offices of Dr Hal Puthoff and listened to the American's descriptions of a near-identical scientific process. *All you had to do – somehow – was perturb the zero-point energy field around an object and, hey presto, it would take off.*

Long before the term had ever been coined, Schauberger had been describing the interaction of his machines with vacuum energy – the zero-point energy field.

Part of the contribution to the craft's ascent, Coats explained, was due to a quasi-aerodynamic phenomenon – something I knew as the 'Coanda effect'. But another part came from the expulsion of the densely compressed 'emulsion' of molecules and atoms that had not been 'virtualised' as they passed through the gill-like slits of the craft's compressor blades. This would have produced a glowing bluish-white luminescent discharge akin to ionisation, exactly as described in the Legend.

I closed the book and reached for the large manilla envelope that Joerg Schauberger had given me before I left his home.

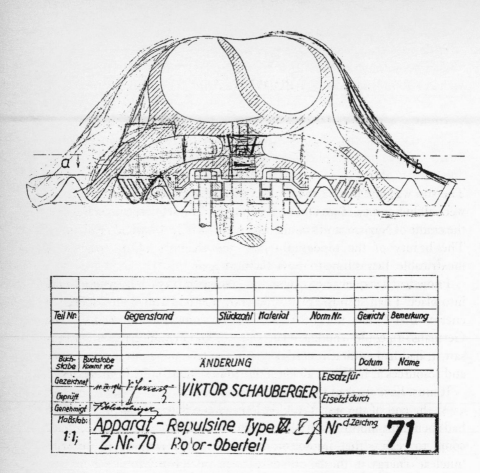

Teil Nr.		Gegenstand		Stückzahl	Material	Norm Nr.	Gewicht	Bemerkung
Buch-stabe	Buchstabe Kommt vor		ÄNDERUNG				Datum	Name
Gezeichnet		*V. Jerieng*	VIKTOR SCHAUBERGER			Ersatz für		
Geprüft						Ersetzt durch		
Genehmigt								
Maßstab: 1:1;		Apparat - Repulsine Type Ⅶ. Ⅴ. Ⅹ Z.Nr. 70 Rotor-Oberteil					Nr. d.Zeichng.	71

'Drawing No. 71', signed off by Schauberger at Mauthausen, showed
the simple innards of the *Schaufelentwurf*, the design for the shovel-
intake, that had bolted onto the top of the craft. I pressed the drawing flat,
amazed that a thing of such organic grace should have come from a place
that had once been the embodiment of evil. I thought, too, of the effect
this must have had on the fragile psyche of the one-time forester, the man
who had wanted to bequeath his grandchildren's generation – my
generation – a limitless source of safe energy, to do good for the world,
but who'd ended up working for the machinery of the Holocaust.

Chapter 22

The sun was just hitting the peaks of the Salzkammergut and the roads were all but empty. Two and a half hours to Munich airport via Bavaria, the cradle of Nazism, and I could box up this whole thing and head home. The beauty of the topography and the shadows of the past were inextricable. It was time to move forward again.

I now had a sequence of events that allowed many of the pieces to fall into place. I had seen the plans of a craft that had drawn on a source of energy that could not be explained by conventional science. The Germans, a people who knew the value of good engineering when they saw it, had had sufficient confidence in the technology to throw money and resources at it. When all else fails, follow the money. It rarely lies.

It must have freaked the Americans out.

At the end of the war, the US Counter Intelligence Corps, the CIC, had spent nine months debriefing Schauberger. They would rapidly have come to realise that implosion technology had nothing to do with 'nuclear' energy in the bomb sense, even though Schauberger himself described the processes at work as 'atomic'.

What they would have learned, however, would have been no less unsettling.

Schauberger believed that his machines were generating an anti-gravity effect whose power potential was unlimited.

The CIC agents would have had little comprehension of the mechanics of the device – why should they? – but somebody must have been spooked by it. Why else did they incarcerate the old man for nine months? Why else did they threaten him?

By the early part of 1946, following the debriefings and their admonition to Schauberger never again to involve himself in 'atomic' science, the CIC would have had in its possession the rudiments of an entirely new propulsion medium – one that 18 months later would have allowed General Nathan Twining, head of the USAAF's Air Materiel Command, to admit to a subordinate in a classified memorandum that it was possible 'within the present U.S. knowledge – provided extensive

223

detailed development is undertaken – to construct a piloted aircraft' with the operating characteristics of a UFO; a craft that seemingly defied the laws of physics.

What had happened to this knowledge?

Part of the answer, I felt sure, lay in some curious synchronicity between the final months of Viktor Schauberger's life and events that had started to unfold at around the same time in quiet corners of the North American aerospace industry.

In the spring of 1958, Schauberger, by now 72 years old and in poor health, was visited by Karl Gerchsheimer, a German-American acting as an intermediary for a US financier and multi-millionaire called Robert Donner, who had heard of Schauberger's inventions and wanted to develop and implement them in America.

Gerchsheimer, a 55-year-old German living in the USA, was a man with a colourful past. He left Germany in 1922, eventually settling in Texas in 1937. During the war, he appears to have been deeply involved in US counter-espionage activities, ending up as an intelligence agent, almost certainly for the CIC. From the war's end in 1945 to 1950, he was the US civilian property administrator-in-chief in charge of all civil administration, logistics, transport and accommodation for the US army of occupation, and in this role was described as 'the most powerful non-military individual in the US zone'. In 1950, he returned to the US, where he set up a metal fabrication business that ended up doing significant business, funnily enough, with NASA.

It was during the 50s, according to Callum Coats' book, that Gerchsheimer met Donner, the retired former owner of the Donner Steelworks in Philadelphia. Donner, Coats wrote, was a 'patriot who waged constant war against subversive activity in the United States'. Between them, Gerchsheimer and Donner became sold on the idea that there had to be an alternative to the use of explosive forces to generate power and motion – a view that Gerchsheimer, through his NASA work, even took up with the master of rocketry himself, Wernher von Braun.

In 1957, Gerchsheimer came to Schauberger's home in Austria, telling him that there were millions of research dollars waiting for him in the United States and an engineering facility in Sherman, Texas, that was all ready to implement his ideas for safe, clean energy production. All he had to do was cross the Atlantic and make it happen. Schauberger, who was suffering from emphysema and a bad heart, took some persuasion, but eventually agreed, the stipulation being that he would stay just long enough to get things up and running – around three months – before returning home to Austria. His son Walter, a trained physicist and mathematician, would remain on in the USA for a year to turn his ideas into blueprints from which the Americans could put his implosion technology into production.

Throughout this dialogue, it is clear that Schauberger felt that Gerchsheimer had come to him as some kind of representative of the US government. And while Gerchsheimer subsequently denied this, his evident connections with US law enforcement and intelligence agencies, connections that he never hid from the Schaubergers, merely reinforced the view in the inventor's mind that the offer carried official US sanction. This impression was compounded when Viktor and Walter arrived in the States and Gerchsheimer and Donner tapped physicists at the National Atomic Research Laboratories at Brookhaven, Long Island, for expert advice on implosion technology. Seven weeks after the Schaubergers' arrival in Texas, Gerchsheimer instructed them to write up separate reports about implosion, then forwarded them to Brookhaven for analysis.

At first, Viktor Schauberger was happy to go along with the Gerchsheimer/Donner plan because at long last he believed here was something that would allow him to implement his free energy ideas for the benefit of mankind. But as the weeks dragged into months, and negotiations in the high heat of Texas turned sticky, Schauberger's health became ever more fragile and he yearned increasingly to go home.

Gerchsheimer and Donner, meanwhile, had become frustrated at the painfully slow pace with which Schauberger was transcribing his ideas onto paper, and the whole relationship broke down.

Prepared to do anything that would enable him to get home quickly, Schauberger made the cardinal error of signing a document that he had not bothered to translate properly. When, later, he fully absorbed the text of it, he realised that everything he owned – models, sketches, prototypes, reports, even intellectual property – had become the sole possession of the Donner-Gerchsheimer consortium.

The release form also stipulated – just as US intelligence agents had, 12 years earlier – that Schauberger was to commit himself to total silence on the subject of implosion and that any further ideas he might develop were never to be discussed with anyone other than designated US personnel.

It was the last straw. On 25 September, five days after arriving back in Linz, Viktor Schauberger died a broken man.

There was no point looking for any explanation of this tragedy on the US side. Other researchers had petitioned US intelligence agencies under the Freedom of Information Act for papers relating to

Schauberger's period in Texas, but had been blocked by officialdom, which would neither confirm nor deny the existence of such records, always citing the National Security Act of 1947.

In seeking to find something that explained this final sequence of events in Schauberger's life from the Austrian's side, however, I had come across a small, almost throwaway reference in the Schauberger archive that had seemed to act as the trigger for it.

Shortly before the Gerchsheimer approach, Schauberger records that he was contacted by two aircraft companies, one American, the other Canadian, and that they both offered him substantial sums of money – the US company, according to Schauberger, pitched its bid at $3.5 million – for the rights to acquire his propulsion ideas.

Schauberger never referred to either firm by name, but as this period coincided exactly with the rise of John Frost's shadowy Special Projects Group and its design of a top secret Mach 3-4 flying saucer for the US Air Force – Project Y2/Silverbug – the Canadian company could only have been Avro. No other Canadian aerospace firm at that time was remotely engaged in such advanced technology work.

I recalled my long phone conversation with Frost's son and his revelation that his father had once made a secret visit to West Germany in 1953. There, at a 'Canadian/UK government installation', according to declassified documents in Canada, John Frost had met with an engineer who claimed to have worked on a flying saucer project at a site near Prague in 1944–45. Since Frost was probing in all the right areas, he would inevitably have come across Schauberger's work; the engineer whom he debriefed might even have known him personally.

Schauberger rejected both offers because neither company would meet his twin over-arching demands: that his turbine should be used 'for the common good' – i.e. for commercial aviation purposes – and that the deal should be made public.

This period, of course, also coincided with the period in which George S. Trimble of the Martin Aircraft company, backed by his counterparts in other aerospace firms across the USA, was talking up the conquest of gravity and the impact this breakthrough would have on air and space flight. The world, they announced, was poised on the brink of an era in which free, clean energy would be the norm; and one day, they predicted, it would take us to the stars. In 1955, Trimble had set up RIAS – Martin's Research Institute for Advanced Studies – whose charter had been 'to observe phenomena of Nature and to encourage, promote and support investigations in search of underlying knowledge of these phenomena ... to discover fundamental laws ... and to evolve new technical concepts for the improvement and welfare of mankind.'

RIAS' charter could have been Schauberger's own.

And look at what had happened next.

No sooner had people started showing an interest in Schauberger again than a man with clear ties to the US intelligence community turns up on his doorstep, tempts him over to the United States and shuts his operation down – permanently.

In 1956, Trimble and others announce to the world that gravity can be harnessed in the same space of time it took to develop the atomic bomb.

Within two more years, all of the companies that had been waxing to the rooftops about the coming anti-gravity revolution had fallen silent on the subject. By 1960, it was as if none of them had ever even thought of it.

In the meantime, Avro Canada, the only company in the world to have openly admitted to working on a flying saucer – a dog of a thing that has trouble even getting off the ground – has had the rights to the real but highly secret technology, the supersonic models that show phenomenal promise, acquired by the United States. Then that operation is shut down, too.

John Frost, the man who headed up Avro's Special Projects Group, takes its secrets to his grave and George Trimble, many years into his retirement, gets spooked by a seemingly mundane journalistic enquiry 40 years on from these events.

I returned to that central question. What was it about these complementary technologies – the saucer airframe and anti-gravity propulsion – that had caused such extreme and violent reactions in people directly linked to them?

For the next three decades, no one in scientific circles would have dared discuss these twin strands of development for fear of the ridicule it would have brought upon them. The cult of the UFO – a phenomenon that has clearly been 'spun' by the military-intelligence community from time to time – simply made this worse. It was only in the 1990s, as NASA and other scientists began to strive for propulsion methods that far outstrip today's state-of-the-art, that the whole anti-gravity business started to gain currency again outside the black world. But even then, you had to proceed with caution.

But the real mystery was this. If the principles of anti-gravity tech-nology had been available since the war – earlier, perhaps, if T.T. Brown's contribution should be given credit – then why were some of the best maths and science brains on the planet still struggling to develop the theory underpinning it?

It was this question, I knew, that would take me back to the States one last time.

Shortly before I boarded my flight at Munich, I checked my mobile for messages. There were four and Marckus had left three of them. Whatever was on his mind, I knew I was on to something, because, for once, Marckus was chasing me.

With one eye on the departure gate and another on the clock, I called him back.

Even over the bustle of movement in the departure hall and the static of a bad line, I could tell something was definitely up.

'You need to get yourself to Sheffield the middle of next week,' Marckus said, his voice scarcely audible. 'Podkletnov's come out of hiding. He's going to be talking about his gravity shielding experiments to a small group of invitees at the university there – aerospace industry and defence ministry types mainly. It isn't a closed session, so you should be able to get yourself on the list. Go talk to him. Something tells me he knows a lot more than he's letting on.'

An airline official at the desk was looking at her watch and beckoning for my ticket. I was about to sign off, when Marckus stopped me.

'There's something else,' he said.

I showed the girl my ticket and passport and carried on talking while I proceeded to the aircraft.

Marckus started to talk about the experiment down the mine. He'd got my email and now he wanted to talk physics.

'Dan, I'm about to get on the bloody aeroplane. This can wait, can't it?'

'I know what they were trying to do,' he said simply.

I stopped a few paces short of the door of the aircraft.

'I know what the Bell was really about,' he repeated.

My tone softened. 'OK, go ahead. I'm listening.' I was staring out of a small window at the end of the jetway. The rain was beating down on the wings of the aircraft. I could see passengers settling into their newspapers through the fuselage windows.

'They were trying to generate a torsion field.'

'What is a torsion field?'

'*Laternentrager* means "lantern-holder". But it's the second codename that's the giveaway. *Chronos*. You know what it means, don't you?'

'Yes, Dan. I know what it means. What is a torsion field? What does it do?'

'If you generate a torsion field of sufficient magnitude the theory says you can bend the four dimensions of space around the generator. The

more torsion you generate, the more space you perturb.' He paused
momentarily, long enough for me to make the connection. 'When you
bend space, you also bend time.'

Chronos.

The third experiment in NASA's BPP study and the one that Puthoff
had predicted would be the first to yield a result.

'*Now*, do you understand what they were trying to do?'

I said nothing. It was Marckus who closed the loop.

'They were trying to build a fucking time-machine,' he said.

Chapter 23

Viewed from the back of the lecture theatre, Dr Evgeny Podkletnov, the man who claimed to have nullified gravity, cut a slight, shy figure as he gazed up at his audience, a mixed gathering of academics and defence industry types.

Wearing my Jane's hat, I had managed to get myself invited to the Russian's presentation by Dr Ron Evans of BAE Systems, formerly British Aerospace or plain 'BAe'. It was Evans who singlehandedly ran the giant aerospace and defence company's anti-gravity effort, Project Greenglow. BAE, I was given to understand, had sponsored Podkletnov's trip to the UK.

Evans' name was one of the two I'd scribbled down in my notes after Professor Brian Young's lecture, 'Anti-Gravity: The End of Aerodynamics?', in London in 1991. The other name, of course, had been that of Dr Dan Marckus.

Though Podkletnov's English was immaculate, he talked so softly that I found myself leaning right forward to catch his words, which were a marked by a strong accent. He didn't look like anyone's idea of a rebel.

He was, I supposed, in his mid-to-late 40s and thin of stature. His dark eyes, which had an intensely haunted look about them, radiated seriousness and conviction. I imagined the day his lab assistant's pipe-smoke had hit the wall of the gravity shield he claimed to have created, he had found himself locked on a course against the science mainstream. Its reaction had been so vehement and overwhelming that he had gone into exile for the best part of four years to recover from the shock of it – and, word had it, to refine his expriments.

For a heretic, Podkletnov had acquired for himself a nice, quiet audience in this, his first appearance since the personal trauma of that time. Aside from a handful of graduates and undergraduates from Sheffield University's Department of Electronic and Electrical Engineering, the small group of invitees included observers from the UK Ministry of Defence, Rolls-Royce aero-engines and BAE Systems.

I let my mind wander back to the day I'd ambled, innocent of all knowledge on the subject, into the marbled hallway of the Institution of Mechanical Engineers in the centre of London eight years earlier to hear Professor Young's lecture. Some of the faces in the audience today looked familiar from that evening, though I couldn't exactly say who or why. If I'd known then what I knew now, I wondered if I'd have ended up where I was sitting today. It had been a long haul.

Podkletnov started by recapping the events that had shaped his life and work over the past decade.

By using rotating magnets to spin up his superconducting doughnut-shaped discs to speeds of around 5,000 rpm, he had found that any object placed in the area of influence above them was losing a part of its weight.

Weight reductions of between 0.3 and 2 per cent were found to be easily repeatable, he claimed. Five per cent reductions had been recorded, although not with the same repeatability. 'For a physicist, this is an enormous weight loss, for a practical engineer it is not,' he noted. I thought of the NASA engineers at Huntsville struggling to reduce the cost of space access. For every hundred kilos of rocket, Podkletnov was offering them a launch weight reduction of two kilos.

He was right. It wasn't a whole lot. But then again, gravity shields weren't supposed to exist.

I jotted everything down as best I could, knowing that the better my notes, the better Marckus' efforts at interpreting them for me in plain-talk later. He was reluctant for us both to show in the same place, especially with so many members of the establishment present. Once again, I was acting as his eyes and ears.

The Russian peppered his talk with references to General Relativity and Quantum Theory, the unique properties of superconducting materials and his admiration for other pioneers in his field, among them Hal Puthoff, whose idea that gravity might be a 'zero-point fluctuation force' Podkletnov professed, with characteristic Russian understatement, to find 'rather interesting'.

He started to talk about torsion fields, a term still fresh in my mind from Marckus' analysis of the experiment down the mine. Igor Witkowski, my Polish guide, had also talked of 'vortex compression', a term used by SS police general Jakob Sporrenberg in his deposition to the Polish courts. Witkowski had remarked how this had linked in with Podkletnov's gravity shielding experiments.

'The observed effect,' Podkletnov said, talking about his own tests, 'might be caused by a torsion field excitation of the physical vacuum inside and outside the superconductor.'

The idea came from somewhere far and remote, then steadied long enough for me to begin processing it.

Marckus and I had spoken at length since my return from Austria about the Bell experiment and Schauberger's ideas. There was no doubt at all in his mind that what the Nazis had been trying to do at the Wenceslas Mine constituted an investigation into the arcane properties of space-time. And the fact that they had tested the effect of the torsion field on vegetation, lizards, mice and, God only knew, humans as well, suggested that there had been a practical aspect to the experiment as well. There was evidence, Marckus had told me, that particles appeared to slow down when they entered a torsion field.

'Since the zero-point energy field is composed of billions of tiny fluctuations of energy that pop in and out of existence every split second, relentlessly and infinitely,' he'd explained, 'anything that can mesh with those fluctuations, so the theory goes, can tap into them and extract energy from the field. That's what some people, myself included, believe that a torsion field does.'

A torsion field, he'd continued, was best imagined as a rotating whirlpool. If you created one of these whirlpools, dipped it into the zero-point energy field, the seething mass of latent energy that existed on an almost undetectable level all around us, there was evidence to show that it reacted in an almost magical way by directing the flow of energy. Anything that rotated, even a child's spinning top, Marckus said, was capable of generating a torsion field, albeit a very small one.

'There's evidence of this?' I'd interjected.

Marckus had looked at me and nodded. 'If you imagine the zero-point field as a giant vat of treacle and man-made torsion field generators like the Bell and Schauberger's Repulsine as mix-masters – food-blenders – then you can begin to visualise their effect.'

Some things were better at stirring the zero-point field than others, Marckus had said. Anything that generated an electro-magnetic field was a case in point.

'The Bell would have been radiating like fuck,' according to Marckus, 'generating electromagnetic energy on all frequencies, from radio waves to light – no wonder they'd buried the damned thing so deep.' The fact that the Germans had also filled the rotating cylinders with a mix of different metals was also significant, he believed. If you could get the proportions exactly right you stood an even better chance of interacting with the zero-point field. But it would have been a very hit-and-miss process – an observation that was borne out by Sporrenberg's testimony. Each test had been very short, lasting on average around a minute.

It seemed very much as if the scientists had been trying to 'tune' the Bell, much as you would a radio set.

'Get it right and you've got a very interesting piece of hardware,' Marckus had told me, 'get it wrong and all you've got is some expensive junk.'

Sixty-two scientists shot dead by the SS told both of us that the Bell tests had to have been at least partially successful.

It had been the same with Schauberger's machine. The air molecules whirling around within it had been spun into such a state of super-excitement that something very strange seems to have happened.

Compounded by collisions of electrons and protons – 'the building blocks of atoms', as Callum Coats had called them – the result had been the creation of a vortex – a torsion field – similar to the Bell's, only without the electromagnetic component.

This was no ordinary 'twister' in a three-dimensional sense, but a coupling device to the zero-point field – a pump, if you will – that not only acted as a conduit for its infinite source of energy, the seething mass of fluctuations, but had combined with an aerodynamic lift component and the Coanda Effect to produce levitation.

If, as Puthoff and others were suggesting, gravity and inertia were component forces of the zero-point field, along with electromagnetism, then the vortex/torsion field was also responsible for interacting with the zero-point field's gravity and inertia properties. Tune the machine right and you could manipulate them.

Manipulate the inertia of an object and you removed its resistance to acceleration. Put in space and it would continue to accelerate all the way up to light speed – and maybe even beyond.

Manipulate the local gravity field around an object and you could get it to levitate.

Both of these pathways to 'advanced propulsion' were being explored within NASA's Breakthrough Propulsion Physics initiative.

But here was the truly wild part. The vortex, Marckus said, wasn't a three-dimensional phenomenon or even a fourth-dimensional one. It couldn't be. For a torsion field to be able to interact with gravity and electromagnetism it had to be endowed with attributes that went beyond the three dimensions of left, right, up and down and the fourth-dimensional time field they inhabited; something that the theorists for convenience sake labelled a fifth dimension – hyperspace.

It was here, they said, beyond the vacuum of absolute space, beyond what we now know to be a 'plenum' (the opposite of a vacuum) flooded with zero-point energy, that the binding mechanisms of the universe actually lay.

To date, we know of four of these fundamental forces: gravity, electromagnetism, the strong nuclear force and the weak nuclear force.

If a whirling torsion field with or without an electromagnetic component was binding with gravity to produce a levitational effect – an *anti-gravity* effect – it wasn't doing so in the four dimensions of this world, but somewhere else.

It explained why the Germans had attempted to use a torsion field to act upon the fourth dimension of time. Time, like gravity, the theorists said, was simply another variable stemming from the hyperspace.

'Say,' Marckus had told me, 'to use an extreme example, the Germans had been able to slow time within the area of the Bell's torsion field, this ceramic-lined chamber, to one thousandth the rate at which it was progressing outside and you sat inside the chamber for a year. What you've done is slow time down on the inside, while on the outside it progresses at its normal rate. Step outside the chamber after a year has ticked by on your calendar and you find yourself a thousand years into the future.'

The chances were, Marckus said, the Germans had only been able to generate a tiny time-perturbation – perhaps of the order of a hundredth or a thousandth of one per cent. Its area of influence, mercifully, would also have been very localised. God only knew, Marckus said, what effect a larger-scale manipulation of the space-time matrix would have had otherwise. Science itself could offer few clues.

I believed from what I had seen in the Schauberger archive that anti-gravity was real and that the Germans had thrown considerable resources into cracking the problem.

It struck me then that they wouldn't have pursued a single pathway to anti-gravity, but several. In the same way that the Americans had pursued several different theoretical and applied approaches to the creation of an atomic bomb.

And in the same way that NASA was investigating multiple different routes to breakthrough propulsion physics for taking us to the stars.

In my mind's eye I imaged the strange henge-like edifice next to the power station. Maybe, just maybe, Witkowski was right. Maybe it had been a test-rig of some sort. A test-rig for a highly unconventional engine or a large circular aircraft.

If the Germans had been looking for an anti-gravity effect at the Wenceslas Mine, they would have needed something robust to test the forces generated.

And what of the weirdest fucking variable of all? The gateway to the hyperspace itself?

There was a burst of applause in the lecture theatre. If I'd got the

Russian right, he was saying that gravity shielding – anti-gravity by another name – was a torsion field 'excitation' of the physical vacuum, soaked as it was with zero-point energy.

The Bell, Schauberger's work and Podkletnov's were all linked by torsion fields and torsion fields appeared to be the key to a host of effects science was at a loss to explain; effects that came from somewhere, some other place, that lay beyond the vacuum of nothingness.

In the Q&A session after the formal part of his talk, Podkletnov had made a reference to something that I needed to follow up on. He'd mentioned that during a recent round of experiments he had generated an even more pronounced shielding effect, one that demonstrated a *consistent* weight loss in objects exposed to the beam of five per cent – double his previous best – for short periods of time.

No doubt mindful that NASA had been unable to produce superconductors of the required size, strength and fidelity to duplicate the Russian's experiments, a member of the audience asked him what size discs he'd been using. With a look of evident pride, Podkletnov replied that he'd been using 30 centimetre discs – so strong that the president of Toshiba had been able to stand on one without breaking it.

Half an hour later, during a buffet organised by the university for its unconventional guest and the assortment of government and industry types who'd come to hear him speak, I managed to talk to Podkletnov in a quiet corner of the room.

He must have known that I wanted to pick up on the Japanese angle, because no sooner had I hit him with my business card, than he announced that most regrettably the topic of Japanese sponsorship wasn't for discussion. He had said more than he ought on the subject and that was that.

This in itself, of course, was interesting. But the block on this avenue of enquiry was so abrupt that for a moment it wrong-footed me.

I grappled for some new purchase on the conversation. 'How about aerospace applications?' I asked.

I half-expected Podkletnov to tell me what he had previously told Charles Platt, the American journalist who'd tracked him down in the wilds of Finland. Platt had revealed that Podkletnov was working on a spin-off of the shielding effect that had allowed him to achieve 'impulse reflection' – using gravity waves as a repelling force. These, Podkletnov had told the writer from *Wired*, would one day permit the development of a 'second generation of flying machines', beyond the brute-thrust types we rode on today.

But Podkletnov went off on a different tack altogether.

'I have discovered,' he said slowly, watching and weighing me as he spoke, 'a new and truly bizarre effect.'

'What kind of effect?' I probed. The low-key nature of the Russian's lecture had already signalled that he wasn't in the habit of using hyperbole.

'I'm not sure that I am ready to talk about it publicly yet,' Podkletnov said, smiling awkwardly.

I was about to return to the subject of the Japanese for leverage, but in the end I didn't have to.

'I am, however, happy to share this news with you,' he said. He paused to give the room a quick sweep. There was no one close enough to overhear us.

'If the superconductors are rotated considerably faster than the 5,000 rpm speeds I've been mainly using until now, perhaps five to ten times as fast, the disc experiences so much weight loss that it actually takes off.'

'Have you experimented?' I asked.

'Yes,' he said meaningfully, 'with interesting results.'

It was then that I asked him if he had heard of Viktor Schauberger. It kind of came from nowhere but in actual fact it wasn't such a wild shot in the dark. There was something incredibly familiar about those rotation speeds.

Schauberger had been generating a levitation effect using rotational velocities of 15–20,000 rpm. If these were rapid enough to produce a usable torsion field, no wonder Podkletnov had been getting 'interesting results' from 25–50,000 rpm.

Podkletnov weighed his response before replying.

'You should understand that I come from a family of academics,' he said after a long moment of reflection. 'Both my father and grandfather were scientists. Shortly after the war, my father came to acquire a set of Schauberger's papers. Some time later, when I was old enough to understand them, he showed them to me.'

'Original papers?' I asked.

Podkletnov said nothing, but in itself this told me as much as I needed to know.

When the Red Army entered Vienna, Russian intelligence agents had made a bee-line for Viktor Schauberger's apartment. There, they had found documents and certain component parts that Schauberger had spirited out of Mauthausen and the SS technical school at Vienna-Rosenhugel shortly before the SS transferred him and his design team away from the bombing to the little village of Leonstein.

Afterwards, the Russians had blown up the apartment. But they had evidently taken what they could from it. And for some reason, Podkletnov's father had acquired some of the plunder.

'Perhaps you know that the Germans experimented quite extensively with gyroscopic forces for aircraft during World War Two,' Podkletnov continued.

I wanted to tell him about the evidence I had uncovered in Austria and Poland, to share some data quid pro quo so I could learn more about his father's work, but out of the corner of my eye I could see one of the civil servants from the Ministry of Defence making his way towards us, a plate of sandwiches in one hand, a notepad in the other. A discussion about Podkletnov Senior and German 'gyroscopic aircraft' would have to wait till another time.

On the way to the train station, I called Marckus and told him about the three significant facts I had learned that day. First, that the Japanese were in on the anti-gravity/free-energy act; and Toshiba wasn't routinely in the business of throwing good money at crackpot ideas. Second that Podkletnov was generating a free-floating levitational effect by rotating his superconductors at 30–50,000 rpm. And third, that Evgeny Podkletnov's father had come to acquire original Schauberger documents soon after the end of the Second World War.

'What would Evgeny's dad have been doing with them, Dan? It's a hell of a coincidence, isn't it?'

'Leave it with me,' he said, 'I'll call you back.'

He did so within the hour. He'd been on the Internet, he said, scanning, processing stuff, eventually finding what he was looking for on the IBM patent-server, a voluminous website. I could tell from his voice that this was Marckus' kingdom; the place where he was happiest.

'I grind away at this material in low gear like a Stalin fucking tank, but I always get there in the end.'

'What did you find?' I asked.

'Something . . . curious. More than a coincidence, I believe. It seems that Podkletnov's father filed a master patent in 1978. It's funny, because I damn near quipped to you earlier that he'd probably turn out to be the father of the Moscow sewerage system. Well, guess what? The patent relates to industrial techniques for the continuous pure enamelling of metal pipes. Evgeny's dad, in other words, had come up with a novel way of lining steel water pipes. It tells us two things: that he was a materials scientist, just like his son; and that he knew about water. My guess is that the Podkletnovs would have been quite comfortably off under the old Soviet system.'

'How did he come to acquire the Schauberger papers?'

'Ah, well, let's postulate a little. Russian intelligence agents enter Schauberger's apartment in Vienna at the end of the war and find a bunch of papers, the contents of which they really don't begin to understand. But they know they're important, because they've already tied Schauberger to the SS secret weapons programme. So they bring the documents back to Moscow and look around for someone who can decode them. Most of Schauberger's work, remember, related to machines that used air and water as their driving medium. What was Evgeny's dad? He was a leading academician in the hydro-engineering field – a pipes man. Not exactly the first person you'd think of approaching to do the work had he lived over here, but then again, in Schauberger's field, remember, there weren't any dedicated experts. So, Podkletnov Senior, a water engineer, got the job.'

'You think the Soviets took the technology further?'

'Who knows? The Russians have never been afraid to exploit weird science. Remember the remote viewing programme kicked off by the CIA under Hal Puthoff? The reason the CIA was into it at all was because it picked up on the fact that the Soviets had started to dabble in the psychic spying business. There is ample evidence that the Russians have been tinkering with torsion fields for a long time, too. But my guess is that they never found enough in Schauberger's apartment to make practical use of his theories.'

'But enough for Evgeny's dad to be intrigued by them,' I suggested.

Marckus paused.

'What is it?' I asked.

'Podkletnov's discovery probably wasn't the accident he's cracked it up to be.'

'You mean, when his assistant's pipe-smoke hit the column of gravity-shielded air he was already looking for an effect?'

'That's certainly how it appears now we know what we know about his father.'

'Jesus,' I said, 'if he can do it on a shoestring, anyone can.'

'Right. Scary, isn't it?' He sounded like he meant it, too.

But I was still left staring at the end-wall of the same blind alley in which I'd found myself on the way back from Bad Ischl.

'Here's the problem I'm facing, Dan. If Schauberger and Podkletnov can crack anti-gravity, more or less in their garden sheds, then why can't NASA, for Christ's sake, with billions of dollars at its disposal?'

He thought about it for a long moment, then lightly cleared his throat.

'My best shot, and it doesn't help you much, is that what you and I have learned tells me that there are two kinds of science. The stuff they teach you in college and all the weird shit they don't. The aerospace and defence industry is inherently conservative. It doesn't like change. This . . . knowledge, if you can call it that, is dangerous stuff. It's change with a capital C and it's not easy to get your head around. The aerospace and defence industry says it likes people who think out of the box, because they're the guys who give us the breakthroughs . . . radar, the bomb, stealth and that. But think this far out and they look at you like you're crazy. They might even put you away.'

I thought back to the article that had landed on my desk all those years ago and my failed attempt to interview George S. Trimble.

Only a maverick, an enlightened one at that, would have pushed for the foundation of something as bold as RIAS – the research institute spun out of the Martin Aircraft Company that Trimble had helped to set up in 1955, at exactly the time he was making his public pronouncements about the imminent conquest of gravity.

I liked to think that all those people who'd been talking up anti-gravity in the mid-1950s had been forced into silence by some murky intelligence outfit tasked with guarding the truth about the Germans' achievements during World War Two.

Yet, they might just as easily have been silenced by their own chief executives – people who realised they were talking up the premature death of the aerospace industry itself. A science that made their whole world redundant.

Like white goods, fast-food chains and cars, the aerospace and defence industry of the 21st century has become a global enterprise. Back in the 1950s, though, as a modern entity, one that we would recognise today, it was only just getting going.

The 1950s was the decade that spawned the intercontinental ballistic missile, the Mach 2 jet fighter, guided weapons and transatlantic jet passenger planes. It was a time when the industry needed brilliant individuals – mavericks like Kelly Johnson and Ben Rich of the Skunk Works – to bring about those revolutionary advances.

Maybe that was why one name had kept popping into my head since my return from Austria.

When I had first enquired about Trimble and RIAS at Lockheed Martin, my contacts there had recommended I interview a man who routinely talked about the kinds of things that had once been integral to the RIAS charter.

A man who tended to talk about Nature, not science; a physicist who looked at things quite differently from other people.

Boyd Bushman was a senior scientist for Lockheed Martin's Fort Worth Division, the part of the corporation that turned out F-16 and F-22 fighters for the US Air Force. If you're interested in anti-gravity, then talk to Bushman, they'd said.

Bushman, the guy who'd levitated paperclips on his desk.

I hadn't, because back then Bushman's theories had sounded just a little too weird. And back then, I hadn't begun to accept the fact that anti-gravity and the multi-billion dollar interests of the aerospace industry went together like chalk and cheese.

But it was all a question of perspective. I saw things very differently now from the way I'd seen them then.

What, I wondered at this moment, was an anti-gravity man doing at the world's most prolific manufacturer of good old-fashioned, jet-powered combat aircraft?

If Bushman was talking up anti-gravity in the midst of an industry that didn't want to know, then maybe, just maybe, he could provide the answer to the $64,000 big one; the discrepancy between clear evidence in the real world of the existence of anti-gravity technology and the people at NASA who were, at this very moment, spending ten times that amount scratching around for theory to make it work.

In two weeks' time, I was due on another defence industry junket to the US West Coast. From there, it would be easy to route back to England via Texas.

Journeying into Kammler's kingdom had taught me the value of analysing data in situ. Now, as I prepared for one last trip to the States, I needed something to help me acquire a different kind of mindset.

I considered how, in July 1986, an F-117A Stealth Fighter crashed at dead of night into a desert canyon just outside Bakersfield, California. For me, the crash had always held a special kind of significance, since it was one of those rare occasions when the black world had been flushed into the light.

In 1986, the very existence of the Stealth Fighter was still a classified secret even though the aircraft had been in full squadron service for three years.

What did Bakersfield say about the system that had been set up to protect America's most deeply classified secrets? Had it been a murder site, what would a profiler have deduced from the traces this incident had left in the environment?

Back at my desk, I started sifting through a pile of newspaper cuttings that I'd kept since the time of the crash. Through the cuttings, I had traced the whereabouts of a number of witnesses to the events of that night. One of them, Amelia Lopez, now a local law enforcement officer, had reluctantly promised to act as my guide should I ever find myself in that neck of the woods.

I checked to see that I still had Lopez's numbers. Then I picked up the phone and called the public affairs department of Lockheed Martin, Fort Worth.

Chapter 24

The flight was several hours out of Heathrow, when I reached into my hand-luggage and extracted the book that Marckus had made me promise I'd read before I got to the States. He'd sent it to me ten days earlier, but in the hurly-burly of stories that needed to be written before I disappeared off to America, and in the course of preparations for the week-long trip itself, I simply hadn't had the time to pick it up.

Not only that, but the book, which was called *Blowback – The First Full Account of America's Recruitment of Nazis, and Its Disastrous Effect on Our Domestic and Foreign Policy*, looked rather academic and heavy-going. Even now, as I gazed at the cover, I found myself reluctant to make a start on it, the more so as Marckus never told me what it was about the book that made it essential reading. But I'd read the newspaper, established that there were no movies on the entertainment system worth watching and had even plundered the in-flight magazine for items of interest. With nowhere left to go, I turned to *Blowback* and started reading.

It began promisingly enough, although the territory itself was more than familiar. By Chapter 3, the author, Christopher Simpson, an American, had got his teeth into his country's recruitment programme of German scientists after the war – an operation that had started under the codename 'Overcast'. As early as July 1945, barely two months after the end of the war in Europe, the US Joint Chiefs of Staff gave its authorisation for Overcast to commence. Its remit was to 'exploit . . . chosen rare minds whose continuing intellectual productivity' America wanted to use in the continuing war against the Japanese and the gathering Cold War against the Soviet Union. Under the top-secret project, the JCS directed that 350 specialists, mainly from Germany and Austria, should be brought immediately to the United States. Among the Germans who came to the US at this time were rocket specialists like von Braun and his boss, Army General Walter Dornberger.

I did as I always did with books that held details of Germany's wartime technology effort: I flicked to the index and looked for references to Kammler. More often than not, Kammler never rated a mention, such was the ease with which he had blended into the bland backdrop of the Nazi administrative machine. Here, in *Blowback*, however, Kammler did score a single entry, but only in the context of his relationship with Dornberger at the Nordhausen underground V-2 plant. An arrangement was struck, the book stated, in which the SS, under Kammler, assumed the day-to-day administration of Nordhausen, leaving Dornberger in charge of the rocket technology and the V-2's production schedules.

By 1946, the Pentagon's Joint Intelligence Objectives Agency wanted to recruit even more former Nazi scientists in the gathering stand-off against the Russians, but because US immigration laws forbade entry into the US of any former Nazi Party officials, President Truman was forced to authorise a more extensive effort than Overcast, codenamed Paperclip, on condition that it was kept secret and that only 'nominal' Nazis – opportunistic party-members who had not committed any war-crimes – were to be allowed into the country.

I read on, but the rest of the chapter told me nothing I didn't already know. Simpson's research was excellent, but it was marred by the fact I'd previously read Tom Bower's compelling work, *The Paperclip Conspiracy*. From this, I'd gleaned as much as I needed to absorb about Paperclip: that by September 1947 it had officially been cancelled, but in actuality had been replaced by a new 'denial programme'; one so secret that Truman himself was unaware it existed. Paperclip continued as a 'deep black' programme (as it would be described today), recruiting its last scientist in the mid-1950s. Through Paperclip, hundreds, if not thousands of former Nazi engineers, many of them with highly undesirable backgrounds, were permitted entry to the United States and employed on cutting-edge defence and aerospace projects, not the least of which was NASA's moon programme.

I pressed on, but from Chapter 4 onwards, Simpson's book took a new turn, describing the CIA's recruitment of former Nazi intelligence agents, the spy networks they set in place against the Soviet Union, and the State Department's elaborate efforts to cover up its wholesale recruitment of these people.

I laid the book down and took a pull of my drink. When Simpson had written *Blowback* in the late 1980s, I could see it had been hard-hitting, revelatory stuff. But I couldn't see why Marckus had gone to such lengths to persuade me to read it.

I continued to wonder about this long after I landed in Los Angeles and arrived at my hotel. But given the time difference between California and the UK, it was too late to call Marckus and by the following morning, as I journeyed up the coast to Santa Barbara, the scene of several appointments with a number of defence electronics companies that day, my mind was pre-occupied with other things.

After Santa Barbara, I headed inland to Bakersfield, met up with Amelia Lopez and proceeded to the site where the F-117A Stealth Fighter had crashed in 1986, two years before it was unveiled. That the security surrounding the plane had remained in place after the crash was mainly down to the fast footwork of special Air Force crash-recovery teams, whose policy included sieving the earth up to a thousand yards from the crash site for pieces, then scattering bits of a 60s-vintage Voodoo fighter prior to their departure to throw the curious off the scent.

To me, Bakersfield said two things. It said that the black world would stop at nothing to prevent America's 'quantum leap technologies' from spilling into the public domain until its bright young colonels were satisfied nothing more could be gained by holding the lid of secrecy in place. And two, in scattering bits of Voodoo, an aircraft that hadn't been flown actively for years, it wasn't averse to sending a signal.

When restrictions on the site were lifted, aircraft enthusiasts went into the hills and picked over the site to see if they could find any evidence that for the first time would point to the reality of the Stealth Fighter's existence. When they brought back little pieces of aircraft and had them analysed, they knew exactly who'd been yanking the chain. The pieces of Voodoo were the tell-tale flourish that anonymous artists leave to denote the authorship of their work to cognoscenti of their talent.

It was in the canyon where the Stealth Fighter had gone down that I saw the book again in a flickering moment of clarity. When Amelia Lopez and I were clambering back across the rocks to our cars, the picture steadied and I knew why Marckus had wanted me to read it. The book was the key to everything, not least Marckus' role in all this. His fatalism, humour, cynicism and cunning clicked neatly into place. It was so simple yet overwhelming it almost took my breath away.

Fort Worth owes its very existence to the military, which in 1849 established a small outpost on the Clear Fork of the Trinity River some 30 miles to the west of Dallas.

A quarter of a century later, the coming of the railroad turned what until then had been little more than a cluster of tents and wooden huts into the biggest cattle railhead west of Chicago, a status it retained until shortly after World War Two.

By the 1940s, Fort Worth had begun to profit from other industries aside from cattle and oil, and as you drive out of the city, heading west, it is impossible not to notice Air Force Plant No. 4, a low building a mile long that rises from the Texas plains like an enormous rock escarpment.

Plant 4 had been set up during World War Two to manufacture the B-24 Liberator heavy bomber. In 1944, when output reached its peak, the plant was turning out almost ten aircraft a day; a feat of mass manufacturing that would never be equalled again.

Today, Plant 4 produces the Lockheed Martin F-16 at just a handful of aircraft per month. Lockheed Martin inherited the F-16 programme from General Dynamics, whose aircraft and missile businesses were purchased by what was then simply the Lockheed company in the early 1990s.

GD had, in turn, acquired the Fort Worth operation from Convair, with which it merged in 1954. During World War Two, Convair was known as Consolidated Vultee Aircraft. Originally centred on San Diego, Convair, together with the USAAF, decided to set up the B-24 production line about as far from the Pacific as you could get, as a precaution against bombing by the Japanese.

In 1942, a young engineer called Robert Widmer moved from the San Diego operation to Fort Worth to begin work on the B-36, a massive intercontinental bomber that would make even the B-24 look tiny. It had been specified in 1939 to bomb Germany from the United States in the event that Hitler succeeded in invading Britain. It rolled out of Plant 4 in 1946 and went on to become the backbone of Strategic Air Command during the 1950s.

Bob Widmer spent much of the late 40s and early 50s lobbying to save the B-36 from cancellation before going on to design the B-58, SAC's first supersonic bomber. But by the late 1950s, Widmer had also begun throwing his energies into another project – one that was so advanced that the details would still be beyond the public domain four decades later.

The project was called Kingfish and it emerged out of a requirement by the CIA for a very fast high-altitude reconnaissance aircraft to replace the U-2, which was rapidly running out of places to hide from the Soviets' SA-2 Gudeline surface-to-air missiles in the rarefied air over the Soviet Union.

As chief engineer of Fort Worth at the time of the Kingfish programme and, almost 40 years later, still a consultant for Lockheed Martin Fort Worth, Bob Widmer was someone I very much wanted to meet.

What was so special about Kingfish that it had remained out of the limelight all these years? Questions about this aircraft take on an even more bizarre twist, when you consider the official history of the programme that spawned it.

I was interested in Widmer and Kingfish, because to me it appeared to offer a link – one that had been overlooked by everyone else – to Aurora, the large black triangular aircraft that had supposedly been sighted over the North Sea in 1989.

In August 1959, the CIA's contract for a high-flying, super-fast reconnaissance aircraft went not to GD Convair, but to Kelly Johnson's Skunk Works, which went on to deliver an aircraft called the A-12 under a programme codenamed 'Oxcart'.

The A-12 flew for the first time in 1962 from the USAF's remote test centre at Groom Lake, Nevada, better known today as Area 51. The 15 A-12s that were built conducted three years' worth of operational reconnaissance missions up to the programme's cancellation in 1968. By then, an Air Force version of the plane, the Lockheed SR-71 Blackbird, had entered service.

With a top speed of Mach 3.4, the Blackbird remained the world's fastest operational aircraft until its retirement in the early 1990s. If you exclude the evidence for Aurora, there is, even today, nothing remotely as fast.

In official versions of the Kingfish/A-12 contest (declassified by the CIA only in the early 1990s) the Convair and Lockheed proposals were remarkably evenly matched. Both aircraft were optimised for cruise speeds of Mach 3.2, but the Lockheed proposal pipped its rival's by tiny margins on key parameters such as range and operating altitude. On this evidence, then, it was only right that Lockheed got the deal.

But in the same way, as Marckus observed, that there are two kinds of science, so there are two kinds of history.

In 1995, a small photo of a windtunnel model of Kingfish appeared for the first time in the pages of an aerospace magazine. Scrutiny showed the model to be much more advanced than the paltry details in the caption suggested.

When I made enquiries at Lockheed Martin, under whose historical aegis both of these programmes (Kingfish and the A-12/SR-71) now fell, I discovered that Bob Widmer, Kingfish's designer, was still employed at Fort Worth as a consultant. At 80 years of age, Widmer was busying himself in unmanned combat air vehicle technology. UCAVs – cheap, pilotless fighter planes – are seen very much as the future. It says a great deal about Widmer's skills as a designer that he was still employed at the cutting edge of aerospace and defence engineering almost 60 years after he joined Consolidated; but then Widmer, I was about to learn, was no ordinary engineer. Like Lockheed's Kelly Johnson and other 'engineer-leaders' of the day, he'd helped to shape the modern US aerospace industry.

His official biography describes him as an 'innovator and a maverick' and the 'first engineering leader to totally integrate airframe, propulsion and avionics into a single weapon system, the B-58.'

What it does not say, but what I had gleaned from others, was that he blended these skills with those of a pastor, acting as guide and mentor to his team. In the 1950s, the pressure on GD Convair to outperform its rivals was intense, but the atmosphere inside the plant was informal and shirt-sleeves – almost familial. To many of the younger engineers, Bob Widmer wasn't just a brilliant designer, but someone they could turn to for advice – something of a father-figure, in fact.

We started by talking about Convair's own black ops capability, a special projects facility that stayed in the shadows, unlike Johnson's entity over at Lockheed, which revelled in its status as black world aviation's one-stop shop for special programmes.

'Kelly was the catalyst, the guy who did things differently,' Widmer said, 'but, hey, we had a "skunk works", too; the difference was, we never spoke about it.'

Widmer's slight, sinewy body, so much frailer than it had been in the black and white shots from the 60s, trembled with excitement. His small eyes shone.

As the talk drifted into the early days of the Cold War, a time when GD Convair came into its own as a reconnaissance specialist, I could see that the minder assigned to me from the PR department was beginning to lose interest. An individual I hadn't come across before, he was there as a matter of routine to ensure that I did not encroach upon matters that would put Lockheed Martin in a bad light or compromise national security. But since this was history, history from way back at that, it hardly constituted a threat to either. It was late and the minder looked tired. His eyes were closed, to focus on the conversation more intently perhaps? But if I hadn't known better, I'd have said he was asleep.

In this atmosphere of near-conviviality, I asked Widmer about Kingfish.

At first, he was reluctant to say a great deal, but seeing that I knew some of the details already, he began to open up. As we talked, I could hear the anger in his voice; anger, I thought, that stemmed from the fact that the CIA had chosen Lockheed's plane, not his, for the contract.

This portrait of Kingfish – the one that Widmer now began to talk about – was markedly different from the official CIA version. And here was one element of the mystery. In its final design form, Kingfish had been at least twice as capable as the variant that is recorded in official agency files.

The final iteration of Kingfish was optimised to cruise at 125,000 feet at a speed of Mach 6.25. It had contained a great many 'firsts', including two powerful ramjets – a radical form of powerplant that, to this day, has never been deployed on an operational aircraft – a blended, stealthy airframe shape composed of high-temperature steel and a special heat-resistant substance called 'pyro-ceram' that also acted as a highly effective radar-absorbent material.

Kingfish would have streaked far above the Soviets' air defences, twice as fast as any aircraft the Russians ever consigned to the drawing board; and it would have been all but invisible to radar.

Kind of chilling, too, was the fact that its specification was a near-perfect match for most people's idea of what Aurora should look like – and Widmer had designed the aircraft almost 30 years before rumours of a high-speed Mach 6–8 reconnaissance plane had begun circulating in *JDW* and *Aviation Week*.

That the CIA bought the A-12 instead of Kingfish is, of course, an enormous tribute to the Lockheed project team, which went on to overcome unprecedented design challenges to make the A-12/SR-71 Blackbird a reality. But while Blackbird is now a museum-piece, the finer details of Kingfish, the loser in the CIA contest of 1958–9, will forever remain locked inside Widmer's head.

At the end of the competition, the CIA moved in and ordered all the blueprints, plans and files to be shredded. They also told him never to talk about it again.

I asked what effect this had had upon him.

'At the time, the CIA were in their glory,' he replied. 'I don't want to go into detail about the kind of life I had to lead . . . how I had to conduct myself, my family and so forth.'

'Was it very restrictive?'

Widmer looked over his shoulder at the minder, who must have detected the change in mood. The minder looked at me and I looked at Widmer.

The former chief designer continued. I could see that he was determined to say his piece. 'It was unbelievable. It wasn't just restrictive. It was counter to my conscience and morals and so forth.'

'The whole business of black programme secrecy?'

He nodded and I got the sense he was speaking not so much for himself as the younger members of his team, the people who'd looked to him for guidance. For an aerospace engineer the black world was a regime that ought to be endured only for a small window in a person's career – and even that was a crime, Widmer said. It was not, he added, what he would call a good formula over a long haul.

When I noticed his hands, I saw that they were shaking.

After a pause that was long enough for me to image the mixture of suspicion and draconian security that had once stalked him, he added: 'I mean, the degree they went through in order to hold the secrecy was unbelievable.'

The look in his eyes was so far-off and intense that I was reluctant to intrude further. But I needed to ask him one last question.

'Did it frighten you, ever?' I said.

'Yes,' he replied, his voice scarcely a whisper.

As I left the room, heading under escort for my appointment with Bushman in the cavernous interior of Plant 4, the minder and I said nothing. Perhaps this way, it was easier to pretend that nothing had happened. But as we walked across the shop floor that had consistently churned out combat aircraft for the past 50-odd years and then stepped back into the strip-lit corridors and past endless security-sealed doors, the air-conditioning cold on our faces, I couldn't help but wonder at the senselessness of it all. One look at Widmer, another at his résumé, told you everything you needed to know. His ambition had been to design and build aircraft that made GD Convair the best plane-maker in the business and which protected America from its enemies. In most people's eyes, that made him a patriot, not a security threat.

But the CIA, or whoever the hell had terrified the bejeezus out of him and his family, had insisted on making the point. And what for?

Had Kingfish, the official loser in the CIA's high-speed spyplane contest, gone on as a 'deep black' development programme, emerging years later as Aurora?

Or was the truth more prosaic?

Had Kingfish simply been too expensive to be developed, forcing someone, somewhere to consign its design principles into a black hole of ignorance, the mythic warehouse that Ben Rich had told me about, where no one would ever be able to get at them?

If so, how many other technologies were in that place? How many more Widmers had been terrified into keeping that perpetual oath of silence?

Hundreds. Thousands in all probability.

Whether George S. Trimble had developed anti-gravity in the black or not, as head of advanced programmes at Martin Aircraft, he would have been forced to abide by the same austere security oath. No wonder, then, he'd had second thoughts about our meeting all those years ago. No wonder, my PR contact at Lockheed Martin had relayed the observation that he'd sounded scared.

Trimble must have been scared for the same reasons Widmer had been.

God only knew what they thought might happen to them.

Like an unsinkable ship, the black world had been built up around multiple, layered compartments, each securely sealed. Some of those compartments, it is now clear, had been designed never to be opened again. Ever.

The Kingfish story provided a tantalising glimpse of the contents of one such compartment. By approaching the black world from another direction altogether, I had come to learn of the contents of at least one other.

At the end of the Second World War, via Schauberger's work and, no doubt, a wealth of other captured technology from Germany that has never seen the light of day either, America acquired knowledge of the most dangerous kind.

The Germans' secret weapons programme had yielded a powerplant – a free-energy source, to boot – that had given America the ability to design a radically different form of air vehicle, one that reflected General Twining's assessment to General Schulgen that a 'circular or elliptical' craft that normally made no sound, which could turn on a dime and was capable of extreme acceleration, could be constructed, 'provided extensive, detailed development' was undertaken.

Something of the order of the Manhattan Project that had yielded the atomic bomb.

But there is no evidence that the Pentagon ever initiated such a programme.

The closest thing to it that has reached the light of public account-ability was the recently declassified Project Silverbug, the USAF's development of the supersonic saucers designed by Avro Canada in the mid-late 1950s. But Silverbug, being jet-powered, was conventional compared to what the SS had tried to develop in Czechoslovakia, Lower Silesia and Austria.

If the US agents who debriefed Viktor Schauberger in 1945–46 so much as half-appreciated the ability of his or other devices to cause disturbances, however small, in the fabric of space and time, to unleash untold 'variables' hidden in dimensions that conventional science hadn't begun to see or understand, then someone, somewhere would have taken the view that it should be kept secret for a very long time.

There were other even more disturbing possibilities.

During their time in America, it had dawned on Viktor and Walter Schauberger that the implosion process could be harnessed to create a massive weapon – one that was infinitely more destructive than the hydrogen bomb. More than anything else, they were terrified that their 'hosts' would latch on to the same idea.

I filed this away in the back of my mind as something to put to Marckus.

Bomb or not, it appears that anti-gravity was boxed up and put away, ready for the day when American science could harness it consistently enough to control it. Besides, in the 1940s and 1950s, it wasn't as if the world really needed it. The global economy – and none more than America's – has been reliant on oil and the technology that supports it for more than a century. It is only now that we're having to confront the end of the petro-chemical era and seek alternative power-sources.

I was left with an image of Widmer's hand, shaking with anger, and a cold feeling I hadn't had since Witkowski told me about the murders of the 62 Bell scientists at the hands of the SS. It didn't take a whole lot of introspection to get to the root of it.

For years, I'd laboured under the assumption that it had been the bomb programme that had kick-started America's black world into being. It hadn't, of course. From a security stand-point, the Manhattan Project was a flawed model. It had leaked like a sieve and the Soviets had acquired the most powerful technology on earth, the atom bomb, a scant handful of years after the Americans.

In 1944, when the Joint Chiefs of Staff established their high-level committee to hunt down on America's behalf every military and industrial secret that Nazi Germany possessed, they couldn't have begun to foresee the full consequences of that directive. America, more than any other ally, had acquired a vast array of German-derived technology. Much of it, like von Braun's V-2 rocketry, was highly advanced, but essentially recognisable. Some of it, however – notably the non-conventional science pursued by the SS – came from a different culture altogether.

To master the weapons it had acquired from the Germans, America found that it had to recruit the engineers responsible for them. Some were former Nazis, but many were just scientists, no more no less, who'd simply been doing their job.

The trouble was, it didn't end there. With the help of German-derived science, America's technological lead over the rest of the world accelerated exponentially after the Second World War. But the black world was a low-grade reflection of the system that had been employed to protect the secrets of the *Kammlerstab* within the confines of the Skoda Works.

The state within a state had been transported four thousand miles to the west and somehow, I just knew, Kammler had come with it.

The intuitive feeling I'd experienced all these years in obscure corners of the US aerospace and defence industry had suddenly acquired a face.

Throughout the next interview, I felt the presence of the info-minders like a low-intensity headache. After the near-informality of my time with Bob Widmer, the proximity of three company caretakers in the room gnawed away at my concentration. They were here to watch over the man who could levitate paperclips on his desk.

But now that the interview was over, the minders were looking the other way. They'd done their job and I'd done mine. Boyd Bushman had spilled no secrets either on or off the record and that was what mattered. Nor had he in any way tarnished the reputation of the Lockheed Martin company.

That I had learned nothing of value must have shown on my face, for without warning, Bushman leaned forward and put his hand on my shoulder.

He asked me what was wrong and I told him.

'It's a lonely walk, but a rewarding one,' he said, so quietly that I almost missed it. I looked into his eyes, which were quite blue but for that superficial milkiness that sometimes denotes the onset of old age.

He smiled at me. 'Keep travelling the road and you might just find what you're looking for.'

'What do you mean?' I asked cautiously.

'In all my years with this company, no one has asked me the questions you came here with today.' He paused a moment, then said: 'Here, I want to show you something.'

He produced a video cassette from the folds of his loose-hanging suit and gave me the same smile that had crossed his face a number of times during the interview; usually, I noticed, when he needed a moment or two to think about a particularly awkward question I'd thrown at him. He looked not unlike Walther Matthau, the Hollywood actor, with a kindly, hang-dog expression and thick straight hair to match. Instinctively, I'd liked him. But more to the point, I knew that I could trust him. Unusually for someone who toiled in the heart of the US defence-industrial complex, he looked like someone who was incapable of telling a lie.

Bushman appeared beyond the retirement age of most good man-and-boy company men, which in itself spoke volumes about his ability.

Before I met him, I'd been told that he was a one-off, one of those people who liked to think out of the box – defence jargon for a guy in the trade who looked at things differently. Just how differently, I was about to find out.

The picture on the TV screen steadied and I found myself staring at a scene of such ordinariness I thought for a moment that Bushman had plugged in the wrong tape.

From the graininess of the film and the way the camera trembled, it rapidly became clear that this was some kind of a home-movie. It depicted the interior of a garage or workshop. In the middle of the frame was a saw on a pock-marked wooden work-surface. Beside it was a pair of pliers and a handful of nails.

For a while, we both just stood there watching a picture in which nothing happened. The minders had taken a quick look and apparently decided it was an eccentric end to a quirky afternoon's work, because they went back to their own conversation.

My gaze drifted back to the TV just as one of the nails in the centre of the screen began to twitch and the pliers moved – jumped, rather – a centimeter to the left. Within seconds, everything was shaking, as if rocked by a series of earth tremors. And then, quite suddenly, the nails stood up, like hairs on a cold forearm.

I looked at Bushman, who was studying the screen intently. 'Watch,' he urged, 'this is the best bit.'

One by one, the nails took off and shot off the top of the screen. A moment later, the pliers started to do a crazy nose-down dance across the work-surface and the saw began flapping around like a fish out of water. Then they both went the way of the nails, shooting out of frame towards the ceiling.

Then, the picture switched: same background, different still-life.

Ice-cream was creeping up the sides of a see-through tub. A lump of it broke away and rose towards the ceiling trailing little pearls of cream. Seconds later, the whole tub jumped off the bench.

My scepticism was working overtime. But Bushman read my thoughts. 'It's no fake,' he said. 'This is as real as you or I.' He cast a look over his shoulder at our companions. 'The man responsible is a civilian, nothing to do with the defence business. That he's a genius goes without saying, of course. You needed to see this, because it's confirmation of the *data*.' He emphasised the word as if somehow it would help me understand what I had just seen. 'You must get to see this man. I think, then, you will be closer than you think to the end of your journey.'

'Data? What data?' I looked at him more closely, hoping for more, but he said nothing.

As I was ushered out of the room, the scientist looked down at the thick pack of material he'd handed me before the interview. I hadn't had time to look at it. Now, I wished I had.

That evening, I sat down with Bushman's file in my hotel room. He'd loaded me down with so many papers, I didn't know where to begin or what I was really looking for. There was a stack of photostats relating to the patents he'd filed during a long and illustrious career that had included stints at some of the giants of the post-war US aerospace and defence industry: Hughes Aircraft, Texas Instruments, General Dynamics and Lockheed.

There was his résumé, which noted his top-secret security clearance, as well as copies of scientific papers, stapled neatly at the corner, with exotic-sounding titles that meant little to me.

My eyes swam over pages of complex mathematical equations and algebra sets. I was looking for clues, but seeing nothing that made much sense.

I pressed the 'play' button on my tape-recorder.

From the tenth floor of my Fort Worth hotel room, the lights of the city shimmered in the sizzling 90 degree heat, even though it was a good half an hour after the sun had slipped below the skyline. Inside, the air conditioning was cold enough to prickle my skin.

'. . . I went back to some of the first work on gravity done by Galileo . . .'

Bushman's voice crackled in the darkness, clashing for a moment with that of a CNN reporter on the TV. I adjusted the volume on the tape recorder, hit the fast-forward button and pressed play again.

'So, what I think is: follow the data and log your data very well and don't throw it away because you have a theory. The theory of gravity is just a theory. Einstein improved on the original theories of Newton and he was verified by data – and data gave precedence. No one believed Einstein until the data arrived. Well, our data's arrived . . .'

I scrolled the tape again. As I did so, I spotted something among the collection of papers Bushman had given me. Tucked beneath the patents and company brochure material on weapons technology was a grainy photocopy of a UFO flying low over a straight stretch of desert road. A hand-written caption underneath identified the location as Santa Ana, California, and the date as 1966.

Flicking past it, I found another cluster of photostats taken from the 'Ramayana of Valmiki – Translated from the Original Sanskrit'. A stamp on the title page identified the book as belonging to 'Lockheed/Fort Worth'.

Puzzled, I toggled with the buttons on the tape-recorder and finally found the part of the interview I'd been looking for.

'Nature does not speak English,' Bushman had been telling me. 'Not only that, but if we verbalise it, we're probably approximating, but not telling the truth. Math comes close, but it isn't there either. What Nature tells us is what must be honoured. It has been talking to us on many domains.'

I let the words echo in my head for a moment.

Bushman was a senior scientist for one of the world's biggest defence contractors; a member of the fraternity that I'd looked to for expert comment all my professional life, and he was telling me to trust in what I saw. Follow data, he'd said, not theory. He worked on programmes that lay beyond the cutting edge of known science.

'I can't talk to all the theoreticians,' he'd told me, 'because there don't exist theories where I am.'

For the past ten years, I *had* been following data and the data had shown me the existence of something real. And yet some of the brightest minds on the planet were struggling to produce theory to underpin it. How could there be no theory for effects that were genuinely manifesting themselves in the environment?

What Nature tells us is what must be honoured. It has been talking to us on many domains.

And we'd only understood a fraction of what it had been saying.

If we verbalise it, we're probably approximating, but not telling the truth.

And then it hit me. That's all theory was or ever had been – an approximation of the truth. Sometimes that approximation worked out and held fast, becoming scientific currency; other times it didn't and was superseded by another theory.

In this case, there were so many holes in our knowledge, it had allowed for people like Podkletnov and Schauberger to fall through the gaps, to show us things that science couldn't account for. Because science theory hadn't mapped them yet.

Not outside Nazi Germany, at any rate.

Einstein had followed Newton and data had verified them both. But in this field, as with other fields of scientific endeavour, the data was still coming in.

It rendered NASA's theory, in this particular area, at least, quite meaningless.

I flipped another piece of paper in Bushman's file and saw a man with a mop of unkempt hair staring at me from a grainy photostat. He was standing in the same garage I'd witnessed in the film Bushman had shown me at Plant 4.

A note told me his name and that he lived in Vancouver – a place, if Bushman was right, that one way or another looked like being where the road ran out.

Chapter 25

George Hathaway, a science and engineering Ph.D who ran a consulting firm on 'non-conventional propulsion technology' in Ontario, Canada, offered a suitable staging post between the reality of sorts that exists in London and the world inhabited by the man in the photostat that Bushman had shown me: self-trained scientist and inventor, John Hutchison.

I met Hathaway in the coffee shop of my Toronto hotel and was gratified to note that there was nothing remotely flakey about him. Hathaway, whom I guessed to be in his early 50s, looked like most people's idea of a slightly unconventional nine-to-five scientist: a tall man, dressed in jeans and a tartan shirt, with wire-framed glasses, Bill Hickok moustache and hair over the collar. The briefcase gave him a suitably businesslike air as well; he was someone you could talk turkey with.

Hearing him talk, I began to feel a lot better.

I was booked on a flight that afternoon to Vancouver, but had been filled with doubts about the whole trip ever since my arrival in Canada.

It was three weeks after my visit to Fort Worth and I had flown half-way around the world on my own ticket to meet a man who claimed to be able to levitate objects with equipment he'd salvaged from electrical shops and thrift stores.

I had managed to grab myself two days off work. Two days to get to Vancouver and back via Toronto. The ticket had cost me a small fortune.

Faced with a sizeable hole in my finances, the whole thing suddenly seemed to stretch credulity, even if it had come with an endorsement from Bushman.

Worse, I'd been unable to make any kind of contact with Marckus.

As was his wont, he had vanished off the face of the planet. I'd called, left messages, sent emails, but he wasn't answering. In the end, I had been unable to put off the opportunity to travel any longer.

Hathaway's ability, therefore, to converse in hard and fast science and engineering terms about Hutchison came as an enormous relief. The key point was that he and a colleague, a South African called Alexis Pezarro, had believed in Hutchison sufficiently to back him financially for a considerable chunk of the 1980s. And even though they weren't linked in any business sense anymore, Hathaway still professed to be a fan of his work.

'There are three kinds of people involved in the advanced propulsion field in my experience,' Hathaway had told me over the phone a few weeks earlier. 'There are the professionals, people like Marc Millis at NASA.'

Millis. The guy who ran the space agency's Breakthrough Propulsion Physics programme out of a lonely office at the Glenn Research Center in Cleveland, Ohio. I cast my mind back. It seemed like a lifetime ago.

'Then there are the Tesla-types who have an intuitive grasp of electromagnetics who come in from the ground up,' Hathaway continued. 'And then there's John Hutchison who doesn't fit into either category.'

Hutchison, it was more accurate to say, fitted into no real category at all.

He had been born in 1945 in North Vancouver and had grown up an only child, surrounding himself with the only things that he seemed to show any real empathy for – machines. Reading between the lines, it was clear that he had not fitted in as a kid. He'd dropped out of school at Grade 10 and had gone on to complete an education of sorts with the aid of a private tutor (mirroring, I recalled, the schooling regime that Thomas Townsend Brown had endured 40 years earlier).

Amassing a large collection of machine tools, steam engines, old guns, chemical equipment and electromagnetic gear, Hutchison became infatuated from a young age with the theories of Einstein and Faraday, even though he was not remotely academic. His interest in them stemmed from his intuitive grasp of science in general and of electromagnetism in particular. His great hero was the turn-of-the-century electrical engineering pioneer, Nikola Tesla.

I had done my homework on Tesla, because Bushman's papers had been peppered with references to him.

Tesla was a Serb engineering graduate from Prague University who in 1884 at the age of 28 emigrated to America in the hope of finding work in the then fledgling electricity industry. With nothing more to his name when he arrived in New York than a letter of introduction to the man widely regarded as the father of the electricity industry, Thomas Edison, Tesla rapidly established himself in Edison's eyes as an engineer of great skill.

But while Edison used him primarily as a glorified Mr Fixit to repair and refine his power generators, which were by then beginning to proliferate across the eastern seaboard, Tesla had set his mind on introducing an electricity supply system that was infinitely better than Edison's. If Edison would only use his, Tesla's, method of alternating or AC current instead of direct or DC current, which was notorious for its inability to travel over long distances, Tesla knew he could deliver a system that was far more efficient.

Yet, by the mid-1880s, Edison's entire infrastructure – indeed, his entire business base – revolved around the supply of DC power, whatever its manifest deficiencies. As a result, Tesla left the Edison Electric Company to set up on his own.

From then on, his career was dogged by a series of business disasters, but as an inventor – and the man who gave the world AC current into the bargain – Tesla was without parallel. In 1891, he developed the Tesla Coil, a remarkable invention that remains the basis for radios, televisions and other means of wireless communication.

But it was the wireless transmission of energy that became his great goal and for most of his professional life Tesla dedicated himself to the pursuit of a system that promised to give the world a free supply of energy. It hardly seemed to enter his head that big business, which by the early 20th century was starting to make handsome profits out of electricity, would be resolutely opposed to the idea.

Tesla's wireless energy transmission system was initially based upon technology for sending power through the air, but he soon developed a far more efficient transmission medium: the ground. In a series of experiments in Colorado, Tesla showed that the earth could be adapted from its customary role as an energy sinkhole – a place to dump excess electricity – into a powerful conductor; a giant planet-wide energy transmission system that obviated the need for wires.

Tesla died penniless in 1943, mumbling to the end about fantastic beam weapons that could blast aircraft from the skies or be bounced off the ionosphere to strike at targets anywhere on the globe. One of the wilder theories that still cling to his memory revolves around a supposed experiment to beam some kind of message to the Arctic explorer Robert Peary, who in mid-1908 was making an attempt to reach the North Pole. According to legend, Tesla, who had built a powerful transmitter at Wardencliff on Long Island, beamed the 'message' on 30 June and awaited news from Peary. The explorer saw nothing, but thousands of miles away, in a remote corner of Siberia, a massive explosion equivalent to 15 megatons of TNT devastated 500,000 acres of land centred on a place called Tunguska.

The Tunguska 'incident' is generally ascribed to the impact of a comet fragment or a meteorite, but the absence of evidence for the impact theory has enabled Tesla proponents to maintain the line that it was Tesla's 'death ray' that caused the blast. Certainly, Tesla himself seemed to believe he was responsible, for immediately afterwards he dismantled the 'weapon' and reverted to other pursuits.

What Tesla and Hutchison shared was a complete sense of ease at being around electricity. Tesla, for example, knew it was perfectly safe to pass high voltages through his own body, provided the current was kept low. Using himself as a conductor, he would stage dramatic demonstrations aimed at validating the safety of AC power. In one of them, he would hold a wire from a Tesla Coil in one hand whilst drawing a spark from the fingers of the other to light a lamp. The newspapers of the day would frequently run photographs of him in morning dress, his whole body alive with sparks. It must have seemed like magic, but was nothing more than straightforward evidence of Tesla's deep-seated feel for his subject.

It was the same with Hutchison. By the age of 15, he had already assembled his own laboratory, cramming it with Tesla machines he had constructed himself; signal generation equipment, static generators and a uranium source housed in a copper tube that used a magnetron pulsed by an old rotary spark plug system to transmit microwaves. During the 1970s, Hutchison set up shop in a bigger, more sophisticated laboratory in the basement of a house in Lynn Valley, North Vancouver, which contained racks full of signal generators, low-power radar systems and phasing equipment coupled to refinements of his earlier devices.

One evening in 1979, Hutchison was in his laboratory, sitting amidst the sparks and high voltage effects of his equipment, when he was struck on the shoulder by a piece of metal. He threw it back and it hit him again. This was the ignominious beginning of the Hutchison Effect. In the ensuing months, Hutchison found that by tweaking the settings on the equipment, he could get things – ordinary household objects – to levitate, move horizontally, bend, break and even explode. In the latter case, this portion of the effect came to be known as 'disruption'.

In 1980, Hutchison went into partnership with a company called Pharos Technologies, which had been formed by Hathaway and the South African Pezarro. Pharos Technologies was designed exclusively to promote the Hutchison Effect, which in these early days was billed as a 'lift and disruption' system. This, however, barely begins to describe some of the extraordinary phenomena Hutchison was manifesting in his Lynn Valley basement.

Using only tiny amounts of power – between 400 and 4,000 watts via a 110 AC wall socket – Hutchison found he could raise any kind of material – sheet metal, wood, styrofoam, lead, copper, zinc . . . whatever – from a few ounces to a hundred pounds in weight at distances of up to 80 feet. From a physics standpoint, it was meant to be impossible, of course, but Hutchison was doing it – regularly, if not exactly on demand. Nine times out of ten, nothing would happen. The rest of the time it was like a bad day on the set of the film *Poltergeist*: objects rising slowly from the area of influence in the centre of the equipment and looping back somewhere else; or shooting skyward in a powerful ballistic arc and striking the ceiling, kicked there by a massive energy impulse; or simply levitating and hovering, continuously, for ages.

What was more, Hathaway said, it had been captured on hours of film and videotape and had been demonstrated in front of hundreds of witnesses.

As for what might be producing the Effect, there were many theories. Some said that it was triggered by opposing electromagnetic fields that cancelled each other out, creating a powerful flow of zero-point energy to any object in the zone of influence; others that Hutchison was causing electromagnetic fields to spin or swirl in some unknown way and that this was the trigger for levitation – shades of the torsion fields generated by Schauberger, Podkletnov and the Bell. Another camp maintained that Hutchison was generating the Effect himself by psychokinesis.

What was interesting, Hathaway said, was that Hutchison himself did not know what caused it and he had no control over it once it started. Sometimes the Effect produced levitation, sometimes it caused objects to shred, tear apart or evaporate; other times, it would cause transmutation – the alteration of an object's molecular composition; Hutchison, like an alchemist, could change one metal into another.

'You're shitting me,' I said.

But Hathaway shook his head. 'John's a wild and crazy guy. He lives in his own world and he can't readily express what he's doing in terms an engineer or a physicist might always understand. You should go see him, no question.' He paused, then added: 'A visit to John is definitely a once-in-a-lifetime thing.'

As the Canadian Airlines A320 tracked westwards above the Great Lakes, I sat alone at the back of the aircraft, my mind still beset by doubts, in spite of Hathaway's no-nonsense assessment of the man I was about to meet.

The one thing that kept me going was the germ of a mystery rooted in some hard, incontrovertible fact I'd managed to glean separately from Hutchison and Hathaway: in 1983, a Pentagon team spent money on Hutchison – tens of thousands of dollars over a period of several months – analysing the Effect.

In promoting Hutchison's technology to potential investors – people to exploit its potential as a propulsion or 'disruption' system – Hathaway and Pezarro had come into contact with a unit inside the Pentagon that was prepared to take their claims at face value.

Led by Col John B. Alexander of the US Army's Intelligence and Security Command, INSCOM, the team put up money to move Hutchison's cramped lab at Lynn Valley to a large empty warehouse on the edge of North Vancouver.

The facility was set up under tight security and anything pertaining to it was classified.

The 'INSCOM group' tasked with analysing the Hutchison Effect comprised five people besides Hutchison, Hathaway and Pezarro – two scientists from Los Alamos National Laboratory, a representative of the Office of Naval Research, an Army R&D specialist and Alexander himself.

The presence of the Los Alamos and ONR representatives appeared to be significant, since both of these organisations were driving forces behind the Strategic Defense Initiative. SDI – 'Star Wars', as it had been dubbed by the media – had been the brainchild of Ronald Reagan's science advisers and Reagan had hawked it to the rooftops as the be-all and end-all arms programme of his presidency: a space-based ring of steel comprising sensors, laser weaponry and missiles that would protect the United States from an all-out nuclear strike by the Soviets.

SDI was unveiled in March 1983 and from then until its death later in the decade was heavily staffed by experts from Los Alamos and ONR.

For the life of me, I couldn't think what Hutchison's work had to do with it.

Soon after Hutchison's lab was moved to the large warehouse on the edge of North Vancouver, the experiments began in readiness for the INSCOM visit. The Pentagon's funding was good for four months. Alexander had provided Hutchison, Hathaway and Pezarro with objects on which to test the Hutchison Effect in its new location. One of these was a six by quarter inch molybdenum rod. Molybdenum is an incredibly strong but light metal used in nuclear reactors. It has a 'sag' temperature of 3,700 degrees C. Alexander had deliberately marked the rod so it couldn't be switched. Give it a whirl, he told Hathaway. See what happens.

A few days later, the colonel was contacted by Hathaway and told that the test-runs had been spectacular. The test specimens were dispatched to the INSCOM team for analysis forthwith.

The Effect had snaked the molybdenum rod into an S-shape as if it were a soft metal – lead, tin or copper. There were other successes to report, too. A length of high-carbon steel had shredded at one end and transmuted into lead the other. A piece of PVC plastic had literally disappeared into thin air before the eyes of the evaluators. Bits of wood had become embedded in the middle of pieces of aluminium.

And, of course, things had taken off – all over the laboratory, items had been levitated; some deliberately, others in parts of the lab that were nothing to do with the experiment. The lab at one point had also become self-running, drawing its electricity from God only knew where – since the power was down at the time.

Keep everything exactly as it is, Alexander told him. We're on our way up.

When the INSCOM team got off the plane in Vancouver, everyone was gibbering with excitement about the looming set of tests – everyone that is, except for the more senior of the two Los Alamos scientists, a dour man who'd been given the unaffectionate nickname 'Big Bad Bob' by the other team members.

Bob remained convinced that somehow it was all just a bunch of jiggery-pokery.

Someone, he'd told Alexander, was being had here.

A shimmy through the airframe as the A320 thumped down on the tarmac at Vancouver and here I was, same place as the INSCOM group, different time.

Like them, I was thinking of the encounter to come; and for a reason I couldn't entirely put my finger on, I felt anxious.

I set out from the hotel for Hutchison's apartment 12 hours later.

It was a crisp morning, warm air from the Pacific mixing with the cold air of a blue sky to form a thick layer of cloud in the waterways of the harbour below. I followed Hutchison's directions, turning left on a wide run-down street, eventually spotting what I was looking for: a brown apartment building with an antenna-farm on one of the balconies.

I could see Hutchison in silhouette at one of the windows, scanning the street for my arrival.

He met me at the entrance and we shook hands. He was dressed in jeans and a denim jacket. A strand of thick black hair fell over one lens of his glasses. The other had a crack running through it. He was tall and powerfully built, but his voice was softer than I remembered it on the phone. Through the handshake I could sense he was as restless as I was. We exchanged a few pleasantries, then I followed him inside. A minute later, there was a rattle of keys and tumbling of locks as Hutchison pushed back the door of his apartment.

The door travelled about 18 inches, then stopped abruptly.

'Sorry, it's kind of a squeeze,' he said, forcing his way through the gap and disappearing inside. I followed suit and found myself in a dark, narrow corridor with what looked like crates lining the walls. The gap down the middle was barely wide enough for a child. Hutchison pushed the door shut and somehow managed to squeeze past me, heading for the back.

He pulled a drape off the window and light flooded the room. Outside, the aerials and dishes I'd spotted from the street pointed skyward at crazy angles. It was only when I turned back to study the apartment interior that I realised the crates I'd seen were radios – huge, metallic, grey, lifeless things, some the size of tea-chests. I peered into the kitchen, the bathroom and bedroom and saw more radios.

Every available inch of floor and wall-space, in fact, was filled with radios and radar-screens, kit that had gone out with the Stone Age, complete with dials, valves and analogue read-outs.

It was like stepping into a cross between a military surplus store and Radio Shack circa 1952.

Hutchison lowered his huge frame to the floor and assumed the lotus position on the chocolate-coloured carpet. He had an almost child-like innocence about him. He'd never once asked why I'd come all this way to speak to him and he gave me the impression that only a small part of his mind was here in the room with me.

The equipment, he told me, had been requisitioned from some Fifties-vintage Canadian warships he was helping to dismantle in the harbour. In return for his salvage services, the Canadian military said he could have all the kit he was able to carry away with him. Clearly, they'd underestimated his strength.

Radios, sonar systems and electronic countermeasures devices lined shelves suspended from the ceiling. Hutchison told me he'd consulted the plans of the apartment block and had drilled holes in all the right places.

'There's a ton and half of the stuff above your head, but it's safe enough,' he smiled. I glanced up nervously and noticed a heavy-looking cluster of boxes suspended over me, then asked if this was the equipment that had been used in the INSCOM tests. Hutchison shook his head.

Much of it had been confiscated by the Canadian police and after their intervention, he'd found it hard to carry on, though recently he'd rebuilt the lab and was readying to start operations again.

He showed me an aluminium ingot that looked as if it had blown apart from the inside and a file filled with letterheads I recognised: Boeing and McDonnell Douglas had been but two of Hutchison's backers.

But he was getting ahead of himself and so was I. First things first. 'What happened after the arrival of the INSCOM team?' I asked.

He scrolled back 15 years and told me about the plough.

The building had been unremarkable, another storage facility in the outer suburbs. There was no security, just an eight-foot wire fence and the warehouse's relative obscurity amidst the smoke and grime of Vancouver's industrial quarter. It was not optimum, but guards would only have attracted attention.

The equipment was housed behind a wall of breeze-blocks in the middle of the warehouse. The breeze blocks extended half-way to the roof – around 15 feet.

This was the target area.

Behind the wall was the large Tesla Coil containing the uranium source, a device around four and half feet tall with a doughnut-shaped metal coil on top.

Diagonally across the room, on the other side of the target area, sat a powerful Van de Graaf generator capable of producing a 250,000 volt DC static charge.

The other prominent piece of technology was a three-and-a-half-foot double-ended Tesla Coil, known as the 'dumb-bell', suspended from supports running across the top of the wall.

Between these three main equipment items were an assortment of smaller devices, all connected to each other by multiple coils of wire and cabling.

There were tuning capacitors, high-voltage transmission caps, RF coils and a spark gap that would snap every 40 seconds or so sending an ear-splitting shockwave throughout the building while it was up and running.

Now, though, as the equipment warmed up prior to the test, all it emitted was a low-intensity hum.

Outside the target area lay a pile of unclaimed scrap metal. Leaning against the wall were three old street lights and a spool of wire hawser. Set in front of them sat a rusting horse-drawn plough, three times the weight of a man.

In the flickering light of the monitoring screens members of the team would see their shadows moving on the edge of their vision. Several said it made it feel like there were people there, watching them.

Between the scrap metal and the target area a bank of receivers and monitoring equipment rested on a workbench.

Hutchison, then in his late 30s, sat at the bench, watching monitors and tuning dials.

The Pentagon and DoE evaluation team had flown in from across the border the previous night. The atmosphere was relaxed as it could be – shirt-sleeves, first-name stuff – but everyone was tense. No one knew what to expect; no one, that is, except Big Bad Bob, the crusty old sceptic from the Los Alamos National Laboratory.

What was happening here, Bob told anyone who'd listen, was just a bunch of smoke and mirrors, a waste of fuckin' time, effort and money.

His sidekick, another John, was blessed with a more open mind and, given the nature of the assignment, he expressed the view, though not when Bob was around, that this was no bad thing.

Only a few days earlier, tongues of fire had licked their way up through the concrete floor, a manifestation of the Effect that had not been encountered before.

Everyone except Bob had accepted the Canadians' version of events at face value.

As Hutchison adjusted the monitoring equipment, he could hear the ladder behind him groaning under Bob's considerable bulk. Bob was gazing down on the equipment from the top of the wall. (Hutchison never made clear to me the reasons for the wall. Had it been put there to contain the Effect or possible fall-out from it – an explosion perhaps?) Colonel Alexander, who was pacing in the shadows to the left of the workbench, asked him what he could see.

The equipment was humming and the cameras were turning. The target, a length of steel rod, wasn't moving – wasn't even quivering – so no friggin' surprise there, Bob replied.

Just then, the radio crackled. Bob's deputy was running some checks with special monitoring equipment outside the building. He was 300 feet from the warehouse and picking up some kind of distortion. Something, he said, was happening.

Alexander stepped out of the shadows and studied one of the monitors. He peered at the readings, then glanced at the video image of the steel bar, but it was just as Bob had said: nothing was moving. Nothing had changed.

Suddenly, without anyone touching anything, the ceiling lights switched on and began to glow intensely bright. For a moment, the entire floor area was bathed in incandescent white light. Then the bank of bulbs blew, sending a shower of red-hot filaments and glass onto the target area.

Except for the glow from the monitors, the warehouse was plunged back into darkness. Bob, who was standing a little to the right of Hutchison, started to laugh.

Out of the corner of his eye, over by the scrap area, the Canadian caught the movement, real movement this time, and instinctively ducked his head.

The crash was followed by a cry from Big Bad Bob, then silence.

The plough had sailed across the room at shoulder height and buried itself into the wall close to where he had been standing.

After the INSCOM team departed, Hutchison's work continued. The lab moved again to a different part of Vancouver and people came and went.

This much I know to be true, because Hutchison showed me copies of the reports.

Among Hutchison's visitors at this time was Jack Houck of the McDonnell Douglas aerospace corporation, who in 1985 spent two days analysing the Effect. Houck came away satisfied that no fraudulence had occurred during the experiments he attended, during which, in a subsequent report, he pronounced that 'some very interesting events' had been captured on video-tape. However, he went on, 'some of the biggest events occurred outside the target area. The first evening a gun-barrel and a very heavy (60 lb) brass cylinder were hurled from a shelf in the back corner of the room onto the floor. Simultaneously, on the opposite side of the room toward the back, three other objects were hurled to the ground. One was a heavy aluminium bar (¾ inch by 2½ inch by 12 inch). It was bent by 30 degrees.'

Houck attested to the randomness of the Hutchison Effect and postulated that it might in part be attributable to psychokinesis – that Hutchison, in other words, was either boosting the energy generated by the machinery with his mind or that his mind alone was responsible for the manifestations produced.

These phenomena, Hutchison was now able to report, included time-dilation – pockets of altered space-time within the target area – and, most extraordinary of all, the capacity of the Effect to turn metal ingots transparent. It was as if, momentarily, the ingots were there, but not there; visible in outline yet totally see-through.

It had been Hathaway's dream in the early stages of their work to develop Hutchison's machinery into a 'stationary launch-assist' device: an anti-gravity aerospace platform.

This was based on his initial belief that the inertial properties of the machinery itself appeared to have altered during some of the experiments; this, in addition to the clearly altered weight condition of the objects placed in the target area – objects that I'd seen in the video Bushman had showed me in Texas.

But in 1986, Hutchison and Pharos Technologies went their separate ways, Pezarro and Hathaway having failed to secure the kind of backing for the Effect they'd always hoped for. It all fell apart, Hathaway said, when it was suggested that Hutchison 'was an integral part of the apparatus'; that the Effect, in other words, was unachievable without the presence of Hutchison himself.

For his part, Hutchison vehemently denies this, but the fact that the most spectacular manifestations of the Effect always seemed to occur when he was around is hard to refute. The psychokinesis proponents hold this up as evidence that their theory is the right one; the technologists counter that it is Hutchison's intuitive feel for the machinery, his ability to tune it to the hair's-breadth tolerances required for things to start happening, that makes his presence necessary.

The fact that two aerospace companies – Boeing and McDonnell Douglas – felt it necessary to investigate the Hutchison Effect is at least telling.

Boeing partly funded a series of experiments during the late 1980s. By then, however, Hutchison had become disenchanted with the Canadian scene and in 1989 went on a scouting tour of Europe intent on moving his lab to either Austria or Germany. When he came back to Vancouver at the turn of the decade, he found the Canadian government had confiscated half of his equipment. The Royal Canadian Mounted Police maintained that it constituted an environmental hazard; Hutchison that the Canadians had been leant on by the Americans.

Hutchison's suspicions were compounded when attempts to have the Los Alamos report released under the Freedom of Information Act resulted in failure.

No one in officialdom had been able to locate the files. It looked as if the report had been buried.

Later, Col. Alexander, the head of the INSCOM group, told me that the report had been classified, but was subsequently downgraded to 'confidential'. In the end, it had been 'routinely destroyed'.

I asked Alexander what he felt about it all.

'Oh, it's real all right,' he told me, 'the trouble is, sometimes it works and sometimes, it doesn't. But four out of five of us came away believing.'

Chapter 26

I met Marckus in a country pub close to the haven on the estuary where we'd hooked up at the very beginning. It had been several years since our initial rendezvous, but when I walked in and saw him in the corner, time might as well have stood still. He looked up from his paper, gave me a wave and clicked his fingers for some attention from the bar. Other than a bored-looking girl behind the pumps, there was no one else in the room.

After she'd taken my order, she came over and set my beer down on Marckus' paper. Marckus clucked like an old hen and handed the glass to me. The circle of condensation, I noticed, had partly ringed the word 'doom' on a headline about the forthcoming earth summit. It felt about right. It had been raining solidly for months.

'So you know,' he said with an air of finality. 'Good. I can't stand awkward explanations.'

Before departing for Canada, when I'd been desperate for him to brief me on Hutchison, I'd left messages on his answer-machine telling him I wanted to talk about *Blowback* – the book he had lent me about US recruitment of high-value Nazis. Marckus must have sussed that I'd known then.

'Where have you been?' I asked him.

'I decided to make use of some vacation time that's been owing to me.' He held up his hands. 'I'm sorry. I should have told you.'

'So where did you go?'

He looked down at his beer. 'I've been following in your footsteps, paying a few respects. Bavaria is beautiful this time of year. From the little hotel where I was staying you can see the snow on the mountains. Maybe on a clear day, if you could only see through the wire, you might have glimpsed the same view from Dachau.'

'How many members of your family did you lose?' I asked.

'A couple that I know of. An uncle and aunt on my father's side. Most of me is as English as you are.' He paused. 'I'm sorry I put you through the hoops. It must all seem a little trite, I suppose. But it was important to me that you found things out for yourself.

'When you started out on the path that you did, sooner or later the trail was going to lead you to Germany. I was happy to help you with the science, but it was important that I held back certain kinds of information. It has taken me decades to build the knowledge that you have acquired in just a few years. But if I'd told you the full story from the outset, you'd have written me off as a crank.'

'Do you think the Americans have mastered the technology?' I asked.

He shrugged theatrically. 'We're going in circles again, aren't we? I told you when we first met. I don't know. I know – you know now – that they have access to it, but whether they've *mastered* it or not . . . it depends on what you mean by "mastered".'

He took a sip from his drink and continued. 'When the Americans tripped over this mutant strain of non-linear physics and took it back home with them, they were astute enough to realise that their homegrown scientific talent couldn't handle it. That it was beyond their cultural terms of reference. That's why they recruited so many Germans. The Nazis developed a unique approach to science and engineering quite separate from the rest of the world, because their ideology, unrestrained as it was, supported a wholly different way of doing things. Von Braun's V-2s are a case in point, but so was their understanding of physics. The trouble was, when the Americans took it all home with them they found out, too late, that it came infected with a virus. You take the science on, you take on aspects of the ideology, as well.'

'And the black world?' I prompted.

'Well, there are different shades of black, of course. But what interests me is that the bit that is truly unaccountable is the part of the system that's been exposed longest to the virus. So in time, it will simply destroy itself. That, at least, is my earnest hope.'

Marckus stared out of the window a while, then shook himself like a dog. 'Tell me about Canada.'

I took a pull of my drink and set the glass down. I felt tired. 'There isn't a whole lot to go by. After the Royal Canadian Mounted Police paid him a visit, Hutchison doesn't even have the equipment to do levitations or transmutations anymore.'

He glanced up. 'So, you believe him.'

'I believe that he caused objects to lose their weight. I also believe he made iron bars turn momentarily transparent and I believe that he transmuted metals.' I tried to rub some of the tiredness away, but the alcohol wasn't helping. 'Beyond that, I don't know.'

Marckus fell silent for a few seconds, then he said: 'Do you have any idea how much energy it takes to transmute a metal?'

At the back of my head, I seemed to recall Hathaway talking in terms of a gigawatt or two. I found it difficult to visualise a gigawatt and said as much to Marckus.

I didn't want another physics lesson, but I realised too late I was going to get one.

'At the low end of the scale, Hutchison was using 400 watts to achieve the effects you're talking about,' Marckus said. 'Four hundred watts are the heat and light you get from four light-bulbs.' A kilowatt – a thousand watts – represents the heat you get from a one-bar electric fire. A thousand kilowatts are a megawatt and a thousand megawatts are a terawatt. A thousand terawatts, Marckus told me, were a gigawatt.

'You know now what a gigawatt is?' he said. 'A gigawatt is the equivalent energy release of the Nagasaki bomb. And that's what Hutchison was pulling out of his 110 volt AC wall-socket to turn steel into lead.

'Transmutation is real. The reason most people don't use it to turn lead into gold is that the power requirements cost more than the gold.'

'Why would INSCOM be interested in alchemy?' I asked.

Marckus smiled again, but the look in his eyes was intense. 'Transmutation has other purposes. You can use it to change the rate at which radioactive elements decay or you can turn the whole process around.'

Marckus was putting me through the hoops again.

'Go on,' I told him.

'Think of it.' he said, leaning forward, 'transmutation, if you can do it the Hutchison way, is a cheap method of enriching uranium, for example. There's another thing, too. Remember when you were heading for Puthoff's place and you asked me for a no-brainer on the power-potential of zero-point energy? I gave you the shoebox analogy and how much untapped energy there is in it?'

He swilled the dregs of his beer around, drained them, then set the glass down and looked at me. 'Maybe Hutchison doesn't know it. Release that energy slowly and you've got a safe, clean reactor that can go on pumping out power forever. Speed up the process and make all the right connections and you've got a bomb; one that'll make a thermo-nuke look like a child's squib. No wonder they shredded the report at Los Alamos.'

I was about to remind him that it had merely been 'routinely destroyed', but stopped myself. Standing next to the coffee machine at the Institute of Advanced Studies in Austin, Puthoff had coolly relayed his belief that there was enough energy in my cup to boil the world's oceans many times over. And then I remembered the Schaubergers' fears over the Americans' hijacking of the implosion process – its potential use in the development of a giant bomb.

I could feel my heart in my throat.

'When he does what he does,' Marckus said, 'Hutchison is reaching into a place that you can't see or touch and he's pulling things out we call "phenomena".' He turned his index fingers into quote marks to make the point. 'But things don't happen by magic. Phenomena happen because there are laws of physics we don't understand yet. The fact we haven't found the answer yet is our problem, not Nature's. But we're working on it.' He stared at me and frowned. 'Why the fuck are you smiling?'

'Because you just nailed something down for me. Something that's bugged the hell out of me almost from the very beginning.'

'What's that?'

'The Philadelphia Experiment. A neat piece of disinformation, I think you'll agree, but stuffed full of real elements that make it impossible for people like you or I to conduct a meaningful investigation of anything remotely related to it. I'm thinking of T. T. Brown, electrostatics and stealth, of course, but there was always that one part of the story that didn't seem to fit – the part about the ship slipping into another dimension. Even that, it turns out, has a grain of truth in it.'

Marckus nodded. 'The best disinformation always does.'

Bushman *was* right. Everything had been right there in Hutchison's apartment.

No wonder INSCOM's weird-shit division had shown up at his door.

Sooner or later, someone would succeed in developing a zero-point reactor – a machine that would be able to exploit the fluctuations in the quantum sea as they blinked in and out of existence millions of times every second. It was just a matter of time. 'We are at the stage that the fathers of the atomic bomb were at when they put together their first test nuclear reactor in the early Forties – and look what happened a few years later,' Puthoff had been quoted as saying recently.

I offered Marckus another drink, but he looked at his watch and said that he had to get going. The wind had started to lash against the window, bringing more rain with it. Marckus pulled up the collar of his coat and readied himself to leave.

But I hadn't quite finished with Dan Marckus just yet.

'So what's it to be?' I asked him. 'Are you going to build a bomb or a reactor?'

Marckus made an exaggerated display of busying himself with his coat. He said nothing, so I asked him again.

'I'd like to know what you're going to do with this knowledge, Dan. You know as well as I do that radical technology that destroys existing technology isn't welcome.'

'Let me put it another way,' he said. 'Years ago, a story appeared in one of those pulp American science-fiction rags, *Astounding Science Fiction* I think it was, in which the US military authorities gathered America's best scientific brains around a table and told them to develop and build an anti-gravity machine, because it was known that the Russians had already developed something similar. The twist in the tail, of course, was that the business about the Russians was a total piece of fiction, but not knowing this the scientists went away, put their heads together, and came up with a crude, but workable device.

'The point of the story is that nothing is impossible, once it is known to be practicable. I have something infinitely better than that. Physicists have proven the existence of the zero-point energy field and your data shows that people have actually developed anti-gravity devices that flew. I want you to know that you weren't simply duplicating knowledge that I already had. Let's say I had the big picture and you provided the details. It's the details in this business that are important. What we're faced with here is a simple set of alternatives. Free, clean energy for everyone on the one hand or the biggest fucking bomb you can imagine. I want the good guys to be in the van here, because a day doesn't go by that I don't think about my family and what happened to them under the Nazis.'

Marckus strolled across the car park, heading for a pathway through the woods on the other side. There was something nagging at the back of my head that I'd meant to ask him, but I couldn't think of it. I closed my eyes, but the inspiration wouldn't come. When I opened them again, he had vanished among the trees.

It was then, of course, that it came to me. I'd meant to ask if he had sent the article that had landed on my desk at Jane's all those years ago.

I smiled to myself. That would have been too Machiavellian; even by Marckus' impressive standards. But I would ask him all the same.

Twenty minutes later, I eased my car along the narrow road that ran beside the abandoned radar establishment and parked up close to the jetty where Marckus and I had first met. The rain had passed, leaving the unmistakable smell of spring in its wake. I got out of the car and gazed skyward. High above a checkered layer of cirrocumulus, an aircraft contrail tracked westwards towards America, the silver arrowhead at the point of its creation looking tiny and vulnerable against the deep blue.

For years, I had banked on the presumption that technology would pursue a set course – one that had been pre-ordained ever since someone had had the wit to fashion a wheel out of a piece of flat stone. Now I knew that it didn't have to be that way, that there were short-cuts in the process that would allow us to leap the state-of-the-art in moments of searing inspiration. I thought back to the day that Garry Lyles at NASA had outlined his vision of interstellar travel, the fast trip to Alpha Centauri that he earnestly hoped man would make within the next 100 years.

If we looked in the right places, we might just be making it a lot sooner.

Epilogue

A few final thoughts, the first of them regarding George S. Trimble Jr., the man whose comments on the imminent 'reality' of gravity-control provided the trigger for the investigation behind this book. In 1967, Trimble left the Martin Company for NASA, where he became director of the Advanced Manned Missons Programme. His reasons for not wanting to be interviewed about his activities at Martin Aircraft deep into his retirement remain unclear, although, of course, they may be entirely innocent. As an ironic twist on everything that followed, I recently came across a comment of his: 'The biggest deterrent to scientific progress is a refusal of some people, including scientists, to believe that things that seem amazing can really happen,' the *New York Herald Tribune* quoted him as saying on 22 November 1955. 'I know that if Washington decides that it is vital to our national survival to go where we want and do what we want without having to worry about gravity, we'd find the answer rapidly.'

Enough said.

In October 2000, the US magazine *Popular Mechanics* revealed that Ning Li, the Chinese-American scientist working in an area of endeavour similar to Dr Evgeny Podkletnov, was close to developing an operational machine capable of exerting an attractive or repulsive force on all matter. By using 12-inch disc-shaped superconductors and around a kilowatt of electricity, Li's device would produce a force field, the magazine stated, that would 'effectively neutralize gravity above a 1-ft diameter region extending from the surface of the planet to outer space.' Li left the University of Alabama at Huntsville in mid-2000 to pursue the commercial development of her invention and has kept a low-profile since. While her eye is said to be on civilian spin-offs of this remarkable breakthrough – NASA scientists at the Marshall Space Flight Center believe her 'AC Gravity' machine could be used as a shield for protecting the International Space Station against impacts by small meteorites – it goes almost without saying that it could also be hijacked as a weapon. The same force that can deflect a meteorite or a piece of orbiting space junk can also stop a reconnaissance satellite or a ballistic missile dead in its tracks. Hit a foot-wide gravity-reflecting beam at an orbiting velocity of 17,000 miles per hour and it would be no different to hitting a brick wall, except there would be no evidence to say that you had hit anything. The results, to borrow an expression from science-fiction writer Arthur C. Clarke, would be 'indistinguishable from magic', the ultimate in weapons technology.

Over recent years, a wealth of articles have appeared outlining the view of an emerging breed of physicists that we may indeed reside in a multi-dimensional cosmos. The August 2000 edition of *Scientific American*, for example, generated a mass of copy in newspapers here and in the States on this growing belief – that parallel universes may exist alongside our own and how additional, unseen dimensions would help to unify the fundamental forces of Nature that inhabit our four-dimensional space-time. According to this thinking, much of it wrapped within so-called 'string theory', our universe may simply exist as a membrane floating within a higher dimensional space, with gravity – this impossibly weak, little-understood yet massively influential force – the only one of the four fundamental forces capable of propagating across the dimensions. Over the course of the next decade, experimentation may provide real answers to these tantalising and complex suppositions. For me, it has simply served to show how taboo notions – taboo *science* – can, in time, enter the mainstream of ideas.

As I sit here, picking over cuttings related to these and other areas of research connected with this book, it has dawned on me only now that attitudes *have* changed in the relatively short space of time it has taken me to write it. Broadsheet newspapers from the *New York Times* to the *Sunday Times* and the *Observer* have all carried thought-provoking features over the past few years on zero-point energy, faster-than-light travel and other contentious areas of science. Suddenly, the idea of gravity having an *anti*-gravity component – this heresy that terrified the professional daylights out of me a decade ago – doesn't seem so strange after all.

On that note, I should add that this book is in not intended to be a catch-all explanation for UFOs. Whilst it may go some way towards explaining some of the thousands of sightings that have occurred since the Second World War – many of them documented in official files – the subject is too complex, too multifarious, in my opinion, to conform to a single explanation. Although the data that Boyd Bushman encouraged me to follow to glean the truth about anti-gravity is sufficient for me to reach some definitive conclusions on that subject, it is inconclusive, to my mind at least, on the question of UFOs. For the time being, it must remain a loose end.

Regarding the definitive fate of SS General Dr Hans Kammler, there are rumours that he died in Virginia – some have it as Texas – but it amounts to insubstantial testimony and to date there is no hard evidence to say that he came to the States, let alone died there. The US National Archives remain devoid of any meaningful files on him.

In April 2001, the CIA provided confirmation of just about everything Christopher Simpson wrote in *Blowback* when under Congressional directive it released twenty files on Nazi war criminals recruited by US intelligence at the end of the Second World War. One of these was Emil Augsburg, an SS officer instrumental in drawing up the 'Final Solution to the Jewish Question' at the Wannsee conference in 1941. Another was Gestapo captain Klaus Barbie, the infamous 'Butcher of Lyons'.

Yet to be confirmed at official levels, though likely to be just as true, is the even less salutary story of Gunter Reinemer, an SS lieutenant in charge of death squads at the Treblinka concentration camp. After the war, the CIA gave Reinemer a whole new identity – as a Jewish Holocaust survivor of all people – before despatching him into East Germany as a spy. In 1988, a few days before he was found dead, Reinemer was unmasked by a German financial investigator named Dieter Matschke. 'You have to ask yourself,' Matschke was quoted in the London *Times* on 15 December 2000, 'how many other Reinemers did America spirit to safety?'

If the US recruitment programme did this for Augsburg, Barbie and Reinemer, it would have bent over backwards to accommodate Kammler, keeper of the Third Reich's most exotic military secrets.

Marckus did not place the article on my desk. I checked.

Bibliography

PRINCIPAL PUBLISHED SOURCES:

Books

Above Top Secret, Timothy Good
Blank Check, Tim Wiener
Blunder! How the US Gave Away Nazi Supersecrets to Russia, Tom Agoston
Blowback, Christopher Simpson
Canada's Flying Saucer Projects. The Story of Avro Canada's Secret Projects, Bill Zuk
The Coming Energy Revolution, Jeane Manning
The Dreamland Chronicles, David Darlington
Electrogravitics Systems: Reports on a New Propulsion Methodology, Edited by Thomas Valone MA, PE
The Elegant Universe, Brian Greene
German A/C Industry & Production 1933–45, Ferenc A. Vajda and Peter Dancey
German Jet Genesis, David Masters
German Secret Weapons of World War 2, Rudolf Lusar
The Goebbels Diaries: The Last Days, Edited by Hugh Trevor-Roper
Hirschfeld: The Secret Diary of a U-Boat, Wolfgang Hirschfeld & Geoffrey Brooks
Impossibility: The Limits of Science and The Science of Limits, John D. Barrow
Infiltration: How Heinrich Himmler Schemed To Build An SS Industrial Empire, Albert Speer
Inside The Third Reich, Albert Speer
Intercept But Don't Shoot, Renato Vesco
Jane's Fighting Aircraft of World War Two
The Last Days of Hitler, Hugh Trevor-Roper
The Last Days of the Third Reich, James Lucas

Living Energies, Callum Coats
Lockheed's Skunk Works – The Official History, Jay Miller
The Making of the Atomic Bomb, Richard Rhodes
The Man Who Invented the 20ᵗʰ Century – Nikola Tesla, Forgotten Genius of Electricity, Robert Lomas
The Military History of World War Two, Edited by Barrie Pitt
Most Secret War, R.V. Jones
The Nazi Rocketeers – Dreams of Space and Crimes of War, Dennis Piszkiewicz
The Nuclear Axis, Philip Henshall
The Paperclip Conspiracy, Tom Bower
The Penguin Historical Atlas of the Third Reich, Richard Overy
The Philadelphia Experiment, Charles Berlitz and William Moore
Raise Heaven & Earth. The Story of Martin Marietta People & Their Pioneering Achievements, William B. Harwood
Skunk Works, Ben R. Rich and Leo Janos
Supertajne Bronie Hitlera, Igor Witkowski
The Biography of Thomas Townsend Brown, The Townsend Brown Foundation
Vengeance. Hitler's Nuclear Weapons, Fact or Fiction, Philip Henshall

Articles

"Black World' Engineers, Scientists Encourage Using Highly Classified Technology For Civil Applications', William B. Scott, *Aviation Week*, 9/3/92
'Breaking the Law of Gravity', Charles Platt, *Wired*, 3/98
'Military Power', Bill Gunston OBE, FRAeS, *Air International*, 1/00
'Nordhausen', Karel Mevgry, *After the Battle*, No.101
'Secret Advanced Vehicles Demonstrate Technologies for Military Use', William B. Scott, *Aviation Week*, 1/10/90
'Secrets by the Thousands', C. Lester Walker, *Harper's Magazine*, 10/46
'Taming Gravity', Jim Wilson, *Popular Mechanics*,10/00
'The Universe's Unseen Dimensions', Nima Arkani, Savas Dimopoulos and Georgi Dvali, *Scientific American*, 8/00
'The US Anti-Gravity Squadron', Paul LaViolette, Ph.D., *Electrogravitics Systems* (Op Cit)
'Without Stress or Strain…or Weight', Intel, *Interavia*, 5/56

Reports

Electroaerodynamics In Supersonic Flow, M.S. Cahn and G.M. Andrew, The Northrop Corporation, AIAA 1/68

Electrogravitics Systems: An Examination of Electrostatic Motion, Dynamic Counterbary and Barycentric Control, Aviation Studies (International Limited)

Project LUSTY, US Air Force History Office

Project Winterhaven, The Townsend Brown Foundation

USAF Electric Propulsion Study, Dennis J. Cravens

USAF Report: 21ˢᵗ Century Propulsion Concepts, R.L. Talley, 5/91